ISBN: 978-1-300-73143-6

TEXTBOOKS ON UNIFICATION SCIENCE

PURSUING
THE UNITY OF SCIENCES

by
Dr. Sung-Bae Jin

UNIFICATION THOUGHT INSTITUTE

KOREA • JAPAN • USA

TEXTBOOKS ON UNIFICATION SCIENCE

Publisher:
Dr. Jin Sung-Bae
President, UTI-Korea

Editors:
Dr. Jin Sung-Bae,
Rev. Hideo Oyamada,
Akifumi Otani,
Dr. Jung Chang Choi,
Dr. Thomas Ward

Coordinator:
Mr Takeshi Furuta

Editorial Consultants:
Mr Jong-Sam Lee,
Dr. Claude Perrottet,
Dr. Richard Lewis

BOOKS IN THE SERIES

***Pursuing* the Unity of Sciences**
Dr. Sung-Bae Jin

Unification Physics
Dr. Ching-Ching Chang

***The Unification Sciences:* Mathematics, Physics and Chemistry**
Dr. Richard Lewis

Unification Medical Science
Dr. Shigehiro Suzuki

Unification Perspectives on Peace and Conflict Transformation
Dr. Thomas Ward and Dr. Claude Perrottet

Unification Ethics of True Love
Akifumi Otani

FOREWORD

I would like to propose a few characteristics of the philosophy that can lead world civilization. First, the philosophy to lead world civilization must have a firm moral framework. The progress of society and culture cannot be measured in terms of technological advances or material superiority. Rather, the maturity of a society and culture should be measured by its standard of goodness. Second, the philosophy to lead world civilization must be able to embrace both the civilizations of the East and West and generate a new civilization through their integration. It should be able to harmonize the technical culture based on the analytical and logical thinking of the West and an ethical culture based on the intuitive thinking of the East. Third, the philosophy to lead world civilization must discover God, lost in modern civilization, and adopt absolute values centered on God as its foundation.

Out of an excessive emphasis on human reason, Western civilization has dethroned God replacing Him with a humanist philosophy, even promulgating atheism and materialism. The first condition of a philosophy capable of leading the world is to establish an absolute and God-centered value system. I see that the philosophy best equipped to lead future world civilization is the Unification Thought of the Reverend Sun Myung Moon. Unification Thought is a new system of philosophy that realizes proper order in the relationship amongst God, human beings, and nature. In this way, we come to discover the true position of God, the meaning of human existence, the meaning of human relationships, and the basis for a harmonious order between humankind and nature.

In his keynote speech at the "International Rally for Victory-Over-Communism and Security" held in Korea on December 16, 1985, the Reverend Dr. Moon described the potential of Unification Thought:

> Unification Thought is a powerful key capable of solving any problem, no matter how difficult it may be. When this thought is applied to society, various social problems can be settled. When this Thought is applied to the world, world problems can be realistically solved. And particularly, when it is applied to Communist ideology and the theory of evolution, all the contradictions of Communism and Darwinism are brought to light, and a counterproposal can be brought forth. This Thought presents a new view of life, a new view of the world, a new view of the universe and a new view of God's work in history. It also offers a principle of integration that can bring different religious doc-

trines and philosophies into unity, while preserving their diverse characteristics.

The Unification Thought Institute (UTI) is dedicated to researching the theoretical foundations needed for a new culture of peace. For he forty years since its founding. since its founding on August 20, 1972 in Seoul, Korea, UTI has sought to address the philosophical and social dilemmas that have led to the chaos and breakdown that undermine modern society. The Institute regularly conducts international seminars and symposia where UT scholars introduce the new thought system and work with other interested parties to unify values across academic disciplines that resonate with values that the world's religions and cultures have long held in high esteem.

Unification Thought was advocated by Reverend Dr. Moon, who completed his earthly life and mission on September 3, 2012. This collection of UT textbooks is meant to provide direction for each of the academic fields introduced in this series. These books are published in commemoration of the Foundation Day of "Cheon-Il-Guk" (the Kingdom of Peace and Unity in Heaven and on Earth) February 22, 2013 (January 13, 2013 by the Heavenly Calendar). On this singular occasion in human history, it is a special privilege for the Unification Thought Institute to dedicate these books to Reverend Sun Myung Moon and Mrs. Hak Ja Han, who, through the Will of God and their unprecedented response to it, now stand as the True Parents of humankind.

Sung-Bae Jin, Ph.D.
President of UTI, Korea

CONTENTS

TOWARDS THE FORMATION OF A NEW SCHOOL OF SCIENCE

KEYNOTE SPEECH AT THE 10TH INTERNATIONAL SYMPOSIUM ON UNIFICATION THOUGHT

MARCH 27, 1998, SUNMOON UNIVERSITY, KOREA

Honorable President Kyung-June Lee, renowned scholars, professors of the Sun Moon University, and all home and foreign VIPs!

I would like to express my heartfelt thanks to all distinguished scholars who have come all the way to this distant land of Korea despite their busy schedules, and all those who have made this symposium a great success by participating in it.

The topic of this symposium is "Constructing Unification Thought in the Age of the Global Village". I feel that Unification Thought, which is the founding idea of the Sun Moon University, should be presented as an ideology that ushers in peace in the era of global village, rather than being limited by the ideology of a specific region or that of a certain country.

At the 9th International Unification Thought Symposium held around this time last year at the Sun Moon University, President of UTI Sang Hun Lee emphasized the following point. The reason for quoting his speech is for us to cherish the memory of the late President Sang Hun Lee, who dedicated his entire lifetime to systematizing the thoughts of the Reverend Sun Myung Moon, and to inherit his idea through this 10th Symposium.

"As you know well, when we watch the current global situation, especially when we watch it from the viewpoint of establishing an ideal world, we find that all spheres of culture including politics, economy, society, media, arts and so on, are collapsing. If we let this situation stand, I am afraid, it may be next to impossible to realize the lofty ideals of humanity. I am, at this occasion, very desirous to provide a momentum for overcoming today's confusion by creating a revival movement centered on Unification Thought. I would like to call this movement 'the Renaissance of Godism.' That is to say, I suggest that we revitalize the Unification Movement by bringing the Renaissance of Godism into existence at this occasion."

The 'Godism Renaissance' Movement presented by the President Sang Hun Lee signifies that all scholastic world cultures must be newly established centering on God, not materialism or humanism. God is the truth of all truths and the source of truth. The posture of learning how to love God is the beginning of all forms of learning and the foundation for solving all social problems. We have to create an opportunity for putting absolute value centering on God in the centripetal point in the various forms of sciences, and re-establish science.

This symposium should be the stage for testing and verifying it. Our concern in this symposium is to demonstrate whether the explanation and logic of Unification Thought can indeed point out the limitations of other varieties of sciences. While acknowledging the superiority of the logic and system of Unification Thought itself, some scholars also brought up problems where it may apply and question Unification Thought in the various sciences. By repeating such a symposium, however, not only can the logic of Unification Thought be applied concretely in various fields such as the current thought, politics, economy, natural science, history and pedagogy, but it can also be presented as a definite alternative measure.

I sincerely hope that this symposium does not only end with the exchange of opinions of various scholars, but in applying Unification Thought to your own area of specialization. A turning point for forming and re-constituting new science will be achieved. Though this is not a meeting of many scholars, having a new perspective in science is of paramount importance. When this movement influences the scholarly world peace in the era of global village will come even more quickly.

It is my hope that through this symposium, Unification Thought will be developed into a movement of science. The centripetal point for shaping science is Unification Thought, and we need to tap the possibility of unifying the various thoughts with Unification Thought as the axis.

A special cultural characteristic of the nineteenth and twentieth centuries that is the most worthy of paying close attention to is related to the formation of a school of science. Unification Thought has the specialty of not being able to be prescribed with the mundane features of school since it unifies all sciences with the absolute value system centering on God.

In this sense, this symposium builds up the flow of a new science with Unification Thought as the centripetal point. Hence I think this should be a gathering by which the movement of science wherein a new viewpoint quickens.

The papers to be presented this time are new in terms of their composition and creativity, and each bears the hard work of the presenters. I would like to request all of you to show keen interest in them and hold lively discussions on them. A culture of heated and highly spirited discussions must thrive in order for a school of science to be formed. Through the course of such discussions, some theory or ideology will be substantiated and the 'durability' of that ideology can be solidified. 'Cool-headed' logic and stringent criticism are the life force of the world of science. Even Unification Thought can only be activated through a culture of lively discussions.

Moreover, the meeting of Oriental and Occidental scholars centering on Unification Thought becomes the cornerstone of a peace movement. Political or economic exchanges are limited by the profit to one's country. On the contrary, a meeting in the real sense of the word is only possible when spirit and truth-seeking spirit, and ideology meet.

May this symposium be a meeting of such kind of spirit and of souls, and a genuine movement of peace be realized. The settlement of peace in the era of the global village begins from a truth-seeking spirit. In this sense, cultural exchange and the meeting of ideologies which make Unification Thought their medium have a very significant meaning. When that happens, Unification Thought will have contributed to the settlement of peace in the era of global village. It is my fervent hope that excellent fruits be borne through today's discussions, even within an unprepared environment. I promise all of you a more wonderful and prepared environment in the future. Thank you very much.

THE RISING TIDE OF THE ASIAN CIVILIZATION AND UNIFICATION THOUGHT

KEYNOTE SPEECH AT THE 11TH INTERNATIONAL SYMPOSIUM ON UNIFICATION THOUGHT

NOVEMBER 23, 1999, MANILA HOTEL, MANILA, PHILIPPINES

I am deeply honored and highly delighted to be able to speak at this 11th International Symposium on Unification Thought, which is being held jointly with the 26th International Conference on World Peace. In this keynote address, I would like to speak on the theme of 'The Rising Tide of Asian Civilization and Unification Thought'.

At the end of World War II, the world was divided into two great camps, the free democratic world, led by the United States, and the. Communist world, headed by the Soviet Union. With the demise of the Soviet Union in 1991, however, the monolithic solidarity of rising tide the communist bloc began to dissolve as each of the member nations chose to steer its own course. In the free bloc, regionalism emerged, and gave rise to regional organizations such as the European Community (EC), the Organization of African Unity (OAU), the Association of Southeast Asian Nations (ASEAN), and the Association of Unity of Latin America (AULA). Although the direct driving force of the emergence of these, regional bodies was a pursuit of common political and economic interests in the region, they can be regarded as manifesting the human aspiration for a global civilization and community of world peace.

The reason why I place a special emphasize on the emergence of the Asian civilization here is that we are now faced with the reality where

Western civilization, which has had overwhelming supremacy up to now, is failing to perform its proper function and on the verge of collapse.In the course of its expansion throughout the world, the modern Western civilization self-destruction. The dynamism and creativity this civilization once possessed before it entered the modern age has now lost persuasion and started on a spiral of rapid downfall.

Originally, Western civilization was highly dynamic, had creative vitality, and played a magnificent role on the historical stage. It is true, however, that this civilizations encounter with, numerous other civilizations around the world in the modem age has generated negative, consequences in many respects. Now, 1 would like to point out a few flaws with the modem Western civilization.

Through the two world wars, Mankind, which had firmly believed in scientific reason, became startled at human brutality, and cast doubt on scientific reason as the basis of Western civilization. Furthermore, modem industrialized society, characterized by mass production and mass consumption, has brought a serious distortion to human nature while providing living conveniences. The human nature created by modern materialistic civilization and competitive society is truly terrifying. The modem person every day has to wage a war of all against all [belium omnium contra omnes] in the logic of the survival of the fittest. What is ruling modern mass society is power 'and jungle ethics of the survival of the fittest. This is an outcome of Western dialectical philosophy, which understands the self and others in the relationship of conflict. This way, through the materialistic civilization of the West, modem people have lost sight of their own dignity and worth; the value of human beings has drowned in the culture of materialism, and the loss of the original human value has precipitated human identity crises.

The crisis of modem Western civilization does not stop here. The structure of the modern industrialized society agrees more with utilitarianism, which pursues the greatest happiness for the greatest number, than with the stem moral philosophy of Kant. The inventions of the modem civilization such as automobiles, refrigerators, and laundry machines are bringing conveniences and comfort to modern people, and as these people see things from the standpoint of convenience and comfort, they cannot but be allured and captivated by utilitarianism and hedonism.

Moreover, such social crises in this industrialized society stems from a collapse of the basic and proper order of human relationship. Machine operations surpass human performance, and the overall industrial structure based on machines has humbled humans and has thoroughly degraded them into being mere parts of machines. The organic world view, which understands human beings as the principle of life, has lost persuasion, and a mechanical world view, which construes them in terms of organization in the framework of a machine, has gained supremacy in the modem society.

Therefore, human relationships have stopped being an encounter of spirits, as we are isolated and alienated in the bloodless structure of organizations finding ourselves in the mechanical framework. In short, we have become slaves of the modern materialistic civilization.

Another point I would like to mention with regard to the problems of the modem Western civilization is the crisis of the confusion of values. In this age, view of values are overturning, and the meaning of life is distorted primarily because the relationship between God and human beings has been dismantled. A person who has lost sight of God can only see spiritual barrenness, existential emptiness, and the meaninglessness of life. Only where God is served at the center, can the true and ultimate meaning of human beings and the universe be restored, and the meaning of life and foundation for absolute value found.

Human beings serving as the standard of all values can only give rise to thoroughly relative values, as exemplified in the world view of the sophists of ancient Greece. Then, as we see in the modem society, materialism and worldly greed may be accepted as the standard of value judgments, and atheism and mammonism may proclaim their legitimacy and gain power.

To enumerate some of the philosophies that have had the greatest impact in Western society, we can list of Karl Marx's dialectical materialism, Darwin's evolutionism, positivism of Conte, and so forth. These philosophies have caused a serious values confusion for the moderns as they elevated materialistic values above spiritual ones while distorting and laying aside absolute values centered on God and human dignity.

Earlier, I have pointed out a few crises of modern Western civilization. These crises have become a general situation of the age rather than an isolated or temporary phenomenon. Such symptoms of Western civilization are offshoots of the limitations inherent in it rather than caused by its confrontation with other civilizations. During its progress through antiquity, the Middle Age, and modernity, Western civilization maintained vitality with its own unity and creative energy; however, when it clashed with other civilizations in modern times, its limitations became exposed. Western civilization has become too weak and old to be able to lead a world civilization. Its expansion throughout the world by means of its economic power and imperialism has ironically created an opportunity whereby its inner sickness surfaced into a full-blown disease.

Many people have foretold a shift of the center of civilization from the West to the East. Western civilization, which began with the magnificent Mediterranean civilization of Greece and Rome, developed through the age of the Atlantic civilization and now has reached its limit and is at a standstill. Judging from the westward movement of civilization, Western civilization, after the age of the Atlantic, will be succeeded by a Pacific civilization, in which Western civilization will blossom and yield fruit. The creativity and energy to lead the world civilization lies hidden in the Asian civilization, which is now about to emerge. At this point, I would like to suggest a few characteristics of the philosophy that will lead the world civilization.

First, the philosophy to lead the world civilization must have a firm moral root. The progress of society and culture cannot be measured in terms of technological advance or material superiority; rather, the maturity of a society and culture should be estimated by the standard of morality. When the leaders of a society lose their moral persuasion, the society is dissolved, and just the same pattern applies to the philosophy to lead a world civilization. We can see this in the ethical philosophy of Socrates, which delivered a fatal blow to the relativist and skeptical philosophy of sophists, which had thrived parasitic on the corrupt democracy of Athens. Likewise, a true philosophy should be able to speak for the conscience of the age with moral persuasion in tunes of crisis.

Second, the philosophy to lead a world civilization must be able to embrace both the civilizations of the East and West and make a new creation out of their integration. It should bring harmony to the materialistic civilization of the West and spiritual civilization of the East. Also, it should be able to harmonize the technical culture based on analytic and logical thinking of the West and ethical culture based on the intuitive thinking of the East. While not ignoring the principle of efficiency, the leading philosophy should lay stress on ethics of the heart.

Third, the philosophy to lead the world civilization must discover God, lost in modern civilization, and adopt absolute values centered on God as its foundation. Out of an excessive emphasis on human reason, Western civilization has dethroned God replacing Him with a humanist philosophy, even promulgating atheism and materialism. Human beings detached from God have slipped into self-worship, and their drive for conquering the environment has ravaged the ecosystem and even endangered the very foundation of their life. Freedom apart from God, however, cannot be true freedom.

It is urgent that we teach a new God-centered system of values from the university level. Platos Academy, the first university in history, had a vast system of academic disciplines, performing ceremonies for the God-Muses, teaching mathematics, philosophy, natural science, and so forth for the purpose of the catharsis of the soul. The dialectics, which they adopted as academic methodology, was none other than a process resembling God supreme universal-, that is the ' Idea of Good'. The spirit of the Academy was that all the studies and education find mutual connection in reliance on God. Today's school education, however, has lost the original Academism, being degraded into a field of transmitting knowledge and technological training. It is the first condition of the philosophy leading the world to establish an absolute and God-centered value system.

I see that the philosophy that is best equipped to lead the world civilization of the future is the Unification Thought advocated by the Rev. Sun Myung Moon. The Unification Thought is a new system of philosophy to bring proper order to the relationship between God, human beings, and nature. In this, we come to discover the true position of God, the meaning of human existence, the social meaning of human relationship, and harmonious order of human beings and nature.

Unification Thought, as well as overcoming the limitation of the Western philosophies, re-establishes the importance and meaning of the Eastern philosophies. There, we come to find the sources of Confucianism and Buddhism re-interpreted in light of Unification Thought. I hope that our honored audience here can nurture Unification Thought as a fruit of the Asian philosophy and furthermore develop it as a philosophy to lead the world civilization. Now we must overcome the obstacles to the formation of an Asian, community such as national selfishness, a drive for national supremacy, and nationalistic grudges, uniting in the spirit of cooperation and living for the sake of others. This way, I sincerely hope that in the new millennium God's Will be realized on earth. Thank you.

AN APPRAISAL OF FORMAL LOGIC FROM THE PERSPECTIVE OF UNIFICATION LOGIC: CAN UNIFICATION LOGIC BE A SCHOLASTIC THEORY?

PRESENTED AT THE 11TH INTERNATIONAL SYMPOSIUM ON UNIFICATION THOUGHT

NOVEMBER 23, 1999, MANILA HOTEL, MANILA, PHILIPPINES

I. The basic characteristics of Unification Logic

1.0. Thinking of a human being is not for the sake of thinking itself, but for the realization of the object.

1.0.1. Accordingly, there is directional nature for the realization of object in thinking, and "Heart" behind it.

1.1. The standard of thinking is the logical system of the Original Image.

1.1.,1. Thinking is a part of creativity..

1.1.2. In the Original Image, an object motivated by Heart is established, where Inner Sungsang (reason) and Inner Hyungsang (law) interact give and receive action, and then Logos comes into being. This is the logical system of the Original Image (inner developmental four position foundation) which can be the standard of thinking.

12. In cognition, or thinking, there are three stages: the perception stage, the understanding stage, and the rational stage.

1.2.12 Cognition and thinking are completed by collecting perception elements and inner Original Image from the outside in the stage of understanding. This is called the theory of collating.

122. In the Original Image, there are the Image of the OriginalBeing and the Image of Form which correspond to the content and form of the sensory thought.

1.2.3. The Original Image which is the subjective element of cognition and thinking is Priority.

12.4. As a human being grows, the Original Image increases its clarity and keeps its continuous relationship with empirical elements.

1.2.5. The subject element of cognition and thinking is interest in the object (directivity for the object) and Original Image.

1.3. Thinking in the stage of reason is completed at the same timeas the Logos (a plan) is formed (completed four position base). But in case that the conclusion is not sufficient, the first stage of conclusion is transferred to the inner idea of the next stage and obtain a new conclusion passing through another process of thinking (Developmental four position base). Thus, the continuous processes of thinning take place in the stage of reason until a satisfactory conclusion is formed without having any empirical relations.

1.4 The form of thinking corresponds to the form of existence.

1.4.1. The reflection of the form of the cell on consciousness is the image of the form, and when this image of form endows thinking with a certain rule, it becomes the form of thinking.

1.4.2. The number of the thinking forms (categories) cannot be definitely limited, but is expressed in a special concept to express the characteristics of Unification Thought (ex. being and prime energy, Sungsang and Hyungsang, subject and object, etc.)

L5. The basic law in all stages of cognition and thinking is the law of the give-and -receive action.

1.5.1. Accordingly, the propositions connected with the logic words or the laws of logic in inference correspond to the give and receive action of correlative forms.

1.5.2. In thinking, freedom of reason is freedom to choose various ideas or concepts in Inner Hyungsang.

II. The limitation of the formal logic from the perspective of Unification Thought.

1. General characteristics of the formal logic

Formal logic is, generally, a scholarship to study the right forms and laws in thinking. The formal logic has unique characteristics of 'formal'. Originally, we cannot separate the content and form in our practical thinking, but we abstractly think about these two things separately in theory. Formal logic is said to be a scholarship to study about the generality of the form of thinking rather than the speciality of content of thinking. Historically it came from the theory of analysis of Aristotle which is based for the first time on the dialectical endeavor of Plato, which has been regarded as a systematical theory for almost a thousand years without being criticized. Kant also was so prompted by this Aristotle's theory that he established the theory of analysis and the theory of judgment on the base of that theory. So this theory has taken a secure position in a solid system with a self identity without being changed. Formal logic has established the reason and laws of the general scholarship and tried to meet essential conditions of the theory in general scholarship and possible study in general.

What the formal logic is seeking to describe is the formative substance of cognition derived from the process of scholastic thinking: (1) formal characteristics (2) possible conditions of the truth connected with that. While the logicians carry out the work to actualize the goal of the reasoning approach of their cognitively acting interest, that is, the cognitive movement, they make use of the middle formative substance, that is, categories. This formative substance is constant and always able to be repeated. And it is not presented by the self-subjective action but is presented objectively. That it can be repeated means that it is always possible to reappear in many cognitive actions and it has the possibility of repeating itself to be perceivable there at any time. Since the cognitive formative substance as the category is not derived from the self-action in the formal logic, such cognitive acting interest as [I pursue....] is excluded.

One of the important essences of the cognitive theory in Unification Logic is the interest of subject in object between subject and object. The reason why thinking of human being has direction towards the object is that thinking is not for the sake of thinking itself. Cognition and thinking are necessarily related with the subjective action towards the object to realize its purpose.

Judging from this point of view, the conclusions of the subjective thinking processes dealt by the formal logic are limited in one part. The concept, judgment, inference, and deduction are, to sum up, the forming substance of thinking. Such forming substance has become an object of the objective research as a result coming

from the process of thinking. Formal logic lost the generalization that the true scholarship should have because of taking interest in the subject of the research which is already directed objectively.

Logic should examine equally of both the subjective condition and the condition at the same time as long as it wants to be a theory about the principles of science, that is, as a true scientific theory. In logic, cognitive forming substance in thinking is an important subject, but the subjective condition of the cognitive subject is also an important one. Logic has to obtain a general domain which is essential for science, and it should be a branch of the science dealing with the basic 'structure' to explain the general domain.

Aristotle's categories of the cognition was succeeded to Kant, who revived it corresponding to the forms of judgment. The categorical judgment of Kant is the judgment combined with quantity and quality, in which the quantity is separated into the generic and a particular. The quality is also divided into affirmative and negative. With quantity and quality combined, the four kinds of judgment can be obtained: Universal Affirmative Judgment, Universal Negative Judgment, Particular Affirmative Judgment, and Particular Negative Judgment. There are also some other modalities of judgment which belong to a complex judgment called a

Reference: Categorical Judgment, Disjunctive Judgment, Hypothetical Judgment, and Causal. Judgment. Therefore, as long as the logic proposed by Kant takes its root in the formal logic proposed by Aristotle, it should meet the same obstacle as the formal logic has faced.

2. Verification of the Formal Logic

Natural human beings live in close connection with the world. As the positive science is generally based on such simple experiences, it doesn't provide a basis for the essential possibility of the general scholarship as the foundation of the general scholarship. That is, the positive science, as it is insufficient of understanding about the fundamental insight for the guidance of itself, cannot be a basically solid scholarship. Even though the scientific experience which is needed in the empirical science is different from the simple experience of the common people, it cannot secure the certainty to become a scholarship because it cannot grasp the point of the profound reason for its own realm theoretically at the every moment due to its exclusive attitudes.

Since formal logic as well as the positive science is based on Simple certainty, it has no universal validity to establish a base for scholarship.

Aristotle obtained the set of categories, turning his eyes upon the problems of common life in the process of approaching to the categories by setting up its foundation from the simple experiences about who, what character, how much, when, where, etc. The traditional logic originated from Aristotle has not caused any problems in the world where it appears directly and simply. And it has set up its reason on the base of the simple positivity.

Accordingly it is called the formal logic. The formal logic externally seems to advocate the essential possibility of the general science, but it actually roots in the simple positivity, which can be said to have lost the foundation of the academic reasoning. For the simple positivity cannot become a base of science in a strict sense of the word.

Accordingly, even though the fatal logic tries to set up the foundation of scholarship, it does not succeed in explaining the real meaning nor having the original lawfulness to obtain self justification, either. Furthermore, it has no criteria to lay the cornerstone over the positivism by positive science.

It is in connection to this problem that Kant had a different idea from that of Aristotle when he set up the logic. He set up a sphere of the priori pure reasoning as the base through which experiments, even if they are not empirical, can be obtained. According to Kant, only the a priori thing existing independently from experiment is straightforwardly able to obtain the universality as well as inevitability.

Husserl also thinks that formal logic does not fulfill its function through the logic essence with the priority objectively because of its relating with the dogmatic simplicity. Therefore, he tries to find out the first subject to be done in the true logic from the explanation of the meaning through a theory of cognition.

He also criticizes," formal logic should display its ability as a pure and universal academic theory having its historical function in mind, but it has rather become a specialized subject for itself."

If the function of logic is to set the foundation to set up a scholarship to be a scholarship, it must be said that the foundation of the "formality" in the formal logic is weak. It is because we cannot obtain the positivity and universality of the scholarship itself from the abstracted "formality" in the simple empirical positivism.

The foundation for experiences to be possible cannot begin at the experience itself. Unification Logic recognizes the 'prior' sphere of the subjective condition as the foundation of such positive experiences to be possible and to control all empirical conditions.

The empirical essence from the simple positivity can be recognized as its propriety only through the interaction with the prior condition controlling the essence. Accordingly, from the viewpoint of Unification Logic, the foundation on which a scholarship can come into being cannot begin from only the empirical " positivity" or subjective "priority" either.

Only the cognitive theoretical 'structure' covering both of these two can make the foundation for the establishment of a scholarship. Viewed from this angle, formal logic is an biased logic structure that is not good enough to secure the universal propriety of scholarship.

3. Noetic contents of the formal logic

I would like to point out the fact that the formal logic makes much account of only the formal logic laws, but takes no account of the contents. In the analysis of the pure logic of the formal logic, the reasonableness on whether the contents have meaning or not is not dealt with.

When we decide the truth or falsehood of the judgment (for example, the golden mountain is false) the criterion of the truth or falsehood is related to the coincidence with the contents. The formal logic does not attach great importance to the reasonableness of the object or to the contradiction of the form of judgment.

What the formal logic sets forth as a premise is not the content of the object which is controlled by the judgment form, but the judgment four'. to control it. The hidden premise of the formal logic does not begin from the under structure called as the object but from the judgmental form which is relatively in the upper part. In the formal logic only the fact that the object should be given to us is premised, but the given condition of the object to tell how it is given is not taken account of. It is a weak point of the formal logic to grasp the point only a little because it sets limits to its logical. concept to the formality only. Kant's attitude on the contents of thought is different from that of formal logic. Formal logic abstracts all contents of the rational recognition and handles only the formality of the judgment. On the contrary, Kant's priory logic puts the rational matters in question within its relationship with recognitional object. For Kant the pure rationality regulates, accepts and unites the varieties given by the sensitivity in a regular way.

For Kant the hyle, variety of the cognitive object, is given by the empirical intuition, which is different from that of formal logic that takes no account of the cognitive contents. For him the cognitive contents are given by the experience. But in the formality of the prior domain of Kant the logical inference is its skeleton, which, it is said, accepts the basic position of the formal logic as it is. It is because the last touchstone of the rational critic of Kant is the law of contradiction of the logic.

This character of Kant's philosophy is criticized by Hussel as the result of his misunderstanding the concept of a priori. Kant did not know that there is a priori in the material thinking contents as well as a formal priori.

Hussel thinks that the origin of the judgmental laws and principles of the formal logic is a partial thing and tries to express the meaning of its pure formality from the priori of the academic theory which cannot be grasped from the viewpoint of the formal logic nor can be seen from the natural thought.

Thus, the pure formality is related with the priority of the academic theory. In short, leaving the actual world untouched in parentheses, it lets the logical proprieties of the world come into question and transfer to the logical dimension. That is, it does not take its interest in the thing behind the phenomenon but takes its interest in the logical structure of the phenomenon itself.

The important matter is what function the phenomenon does in the pure logical domain, and how it reveals itself internally in the consciousness. Thus, in a different method from the empirical one, he opens the way to see the logical structure of the phenomenon through the priori or the logical method excluding all empirical things. What comes out through this result is the object of the meaning, which is the sphere of the content of the judgment that has been left undiscovered in the history of the formal logic.

Therefore, he advocates that .we should present the content of the thought as well as the result of thinking in the judgment form in order for the judgment to be the logic of the pure possibility about the judgment itself. He thinks that his logic is the pure logic, for it shows the pure possibility of the logic.

In the viewpoint of Unification Thought the content of thinking is a priori which is already given in the thinking subject. The concept of the priory original forni which this cognitive subject has becomes subjective conditions that make the cognition and, thinking to be possible.

This subjective condition and the material symbols that come from t he object of thinking come to a coordination. Therefore, the mind that is the original from of the subject, that is, the concept and the symbol of the creature is analogous to each other.

As viewed in the formal logic, if only the formality of the judgment is dealt with, paying no regard to the content of the judgment and thought, it cannot become a logic of a fundamental science, which should include all spheres of the scholarship. Kant, dealing with only the formality of thinking in the priori sphere, advocates that the content of the thought comes from experience as a posteriori. Therefore, the priori logic of Kant's cannot be regarded as good enough for a cognitive logic.

From the viewpoint of Unification Thought, the cognition cannot be formed when the content of judgment has its roots in the empirical sphere while the prior sphere provides a basis for the formality of the judgment, for cognition is a judgment.

It is possible to form a recognition when we recognize that the content of a thought is given logically as a subjective condition of the judgment As long as the' recognition is a judgment, the forming of the cognition is not possible until the subject and object of the cognition is compared and verified.

The formal logic deals with various conditional propositions of the forms of logic, but it does not deal with "the given thing" of the object of the judgment. Therefore the formal logic cannot be said to meet conditions of the hypothesis in the true formal logic.

4. Judgmental form of the formal logic

Aristotle's formal logic is a judgmental analysis in the sense that it asserts whether it is affirmative or negative in the sphere of judgment. Judgment in the formal logic is a general and pure formal concept, that is, a general judgment. For example, it is a general judgment of [S is P1, or [If S is P, Q is R]. When we choose [S is P] as a basic form, [Sp is Q], or [Spq is R] is a subordination or variation of it.

When we deal with the sphere of the basic form of the judgment [S is P] in formal logic, we can grasp only the form of the negative condition through the structure of the form. Whether it is true or false cannot be decided in the condition explained by the precondition of [S is P]. Therefore, the formal logic lays stress only upon the truth value of the judgment that is decided by the relationships between the structure of judgement and another judgment. And S in the judgment of [S is P1 in the formal logic is said to be an ideograph of a subject.

In the formal analysis of the formal logic, the basic element S seems to be thought as something general (Etwas--Uberhapt) because of its unique characteristics in the formalization of the analysis. This kind of formalization is a characteristic of formal logic and also a limitation of the formal logic.

In the viewpoint of Unification Logic, human logical structure is connected with the existing structure which corresponds to it. It is only in this correlation that normal thinking can take place. Therefore, Unification Logic criticizes the formal logical structure of the formal logic which excludes the existing relationship.

Formal logic deals with only the logic structure, ignoring its relationship. In short, formal. logic explains only one part of the study of logic and fails to see the rest of it. You may describe the structure of the formal logic very deeply, but you cannot grasp the real object of the logic or real contents.

The basic body S in the form of the judgment cannot be recognized only through the law of formal logic, but can be explained only when you come over the formal conditions. Total content of the judgment which [S is P] indicates cannot be identified until the form of judgment has an existing relationship with the object.

Clearly, the Unification Logic does not generalize the basic body S into something as we see in the formal logic. All judgements we can think about do not turn toward "something general", but have a special relationship with special objects. It is because an inevitable relationship is formed between man, subject in recognition, and the whole creation, object in recognition. Thinking and recognition of man should not leave any abstract vacant thinking as we see in formal logic. On the contrary, they should have an inevitable relationship with concrete individuals.

All conclusions therefore can go back to the final conclusion which is related with individual objects. Human thinking of course can set up any abstract object in the rational stage and carry out the work of the pure formalization. But when the matter of lucidity of the logic is inquired into, the logical structure of thinking must finally be faced to the structure of existence.

All kinds of functions in the formalization of pure mathematics or in a theoretical study or in the abstraction of arts are also based on the emotional and rational stage in the end. Therefore, all kinds of formal judgment types are said to have their clearness when they are faced to a concrete individual being or situation.

III. Building an academic theory of Unification Logic

1 . Teleological argument of Unification Theology

Logic is a field of study dealing with the laws of human thinking, but it should not confine human thinking within thinking itself. Thinking action of human beings following the thinking rules manifests itself as a way to realize the complex object of the human life. Therefore it is said to be an intellectual ability which is one of the total creative abilities which the original human nature has. Thus Unification Logic explains the basic logic in a total relationship including the sphere of human activities without setting limits to the thinking abilities. That is to say, it does not set limits to the view of cognitive logic but lays its essential foundation on the total relationships of a being. In the process of thinking there is a direction to realize the object of a being, and in the foundation of the direction there has been an object of the being centering on Heart, Shimjong. This direction is the essential reason which works consciously or unconsciously in the thinking process.

From this point, Professor N.A. Von Juven said about the title of Unification Thought from the phenomenological viewpoint that "in the phenomenology, such a relationship between subject and object as is distinguished between the function of direction (desire) and the action of direction (judgment) is related to the relationship of the existence rather than to be restricted to the side of epistemology."

From the viewpoint of the relationship of existence, phenomenology and Unification Thought can be studied in the same chart. UT deals with the object of Creation, while phenomenology begins with the prior-self on the basis of Noesis and Noema structure of inner consciousness.

Moreover, the logical structure of thinking to attain the object of the Creation is related to the cognitive structure and the structure of existence and the structure of dominion on the whole. In order words, since the logic is comprehensively related to the practical aspect of epistemology, ontology, ethics, pedagogy, etc., the structure of logic, as mentioned above from Unification Thought, should be studied inclusively through the connections of various related fields.

2. Logical structure of Unification Logic

The logical structure of human thinking resembles the logical structure of the Original Image in principle. The direction to actualize an object which is based on the Heart let a plan be shaped as a thinking result through the mutual. relationship and unity of the two elements, that is, the directional structure of the rational faculty and the intention, its corresponding concept. The process forming the logical structure of this four position base is not molded by experience but by priority. But it does not mean that the logical structure is made up without experience. Without experience of daily life, we cannot explain the interaction of the logical structure and the practical structure.

When we say that cognitive logical structure is the prior construct, it does not mean that the structure is only a conditional proposition as it is explained in the prior thinking mode of Kant. It means that the logical structure is not the established formal conditional proposition, but the logical structure in which we can explain the content of our experience through the total interaction between the experience of the cognitive object and the physiological process of the body.

Accordingly, Unification Thought has prior logical structure with the relationship of experience at the same time, which is distinguishable from the prior thinking form without experience of Kant. Therefore, the logical structure of Unification Thought is not modeled by experience, or prior formal conditional proposition to prescribe the content of experience. It consists of the prior structure to grasp the content of experience through experience. From the view point of this, Unification logic is distinguishable from the formal logic which consistently prescribes the form for language.

Logical structure should be explained in connection with cognition structure, for the logical structure of cognition is not a simple formal middle concept. The logical structure of Unification Thought is rather similar to Husserl's priori logic in the viewpoint that he admits a priori materials from an experienced world. That is why Hussel criticizes the traditional formal logic as well as Kant's priori logic. Husserl tries to explain in his later writings about a priori logical structure by dealing with the given materials from the world where they live, that is, Noema which is set as an object to be described in the relationship with the directivity for an object.

On the other hand, the process in which any concept is formed in Unification Thought has a possibility to include the experienced content It is not because of the experience itself, but because of the rational faculty in man to realize the ideal world of Creation. Thus, the formative stuff like a concept or idea which is formed in the transcendental cognition becomes the standard of cognition, that is, original form. And it has some relationships with the symbols which are formed from the object of cognition. The logical structure which is formed as the priori experience should ' have a cognitive logical structure for its practical experienced relationship with the practical world in the process of cognition.

3. The basic thinking form of Unification Logic

The focus of the formal logic is not on the content of object to be regulated by the judgmental form, but on the judgmental form to regulate the content of the object. We pointed already that, in formal logic, there is only one proposition that the object should be given to us without any consideration about the content of the datum itself, that is, how it is given to us.

In the modem age, Kant established 12 categories in judgment, and each category comes out of the judgment which gives shape to the logical form. As for Kant, it would not be possible to explain the experienced world of object, namely, an objective experience that is antagonistic to simple occurrence of impression.

Kant calls this demonstration 'the priori deduction of the categories'. The experienced recognition in this demonstration has no specific sphere for its object, but aims at the necessary condition of all experienced cognition.

Husserl's judgment on both formal logic and Kant's priori logic is that they are only used within the limitation of the form of logical concept. Accordingly, Mussel, replaced the form of thinking with the content of thinking. From the viewpoint of Unification Thought, the problem of form and content in logic has been a point of dispute because they have insisted on one side of the content leaning to special parts.

Priori concept of Original Image in Unification Thought is the concept including the form of thinking as well as the content of thinking. Therefore, the form of thinking means that thinking has a certain tendency when it is controlled by the form of physiological existence reflecting human subconsciousness. It can be said that the statue of fowl becomes the shape of form, that is, the form of thinking or the form of understanding. The form of thinking derived from such process has priori form of logic to be recognized through verification with existence form of the object of cognition.

In Unification Logic, the form of thinking or the category of thinking is not a standard form derived from priority as it is in Kant, but it is the proper form human has in his original image as the standard to judge existence form of the object. Unification Logic takes the traditional Kant-like 12 categories as secondary categories, and has a different view in number, that is, it can set more categories than 12.

Unification Logic also checks out the blind point in the Husserl's priori logic. According to Husserl, what clearly appears upon the emotional intuition is not the medium concept of the form or category of thinking but the given object only. In Unification Thought, such cognitive content is checked up with the relationship of the emotional content itself from outside. Unification Thought also recognizes the form of understanding in its logical structure, which was excluded by Husserl.

Viewed from many angles as mentioned above, Unification Logic can explicate both the pure language form in formal logic and Kant's priori logic and Husserl's structure of priori logic in the gross. In other words, Unification Logic has such comprehension and logical structure that can explain all weakness of the structures in the established logic. In this paper I was not able to criticize about the priori logic of Kant and Husserl because of the limited space. I have examined the possibility for Unification Logic to be a scholastic theory.

Notes and References

1. E. Husserl, *Die Idee der Phanomenologie*, 1958.
2. E. Husserl, *Formale and Transzendentale Logik,* Haaf, 1974.
3. E. Husserl, *Logische Untersuchunchungen*, Haag, 1975.
4. Coplesten, *A History of Philosophy*, ed., Ivols, Maryland, 1960.
5. I. Kant, *Kritik der Reinen Vernunft*, Reclam Ibliothek.
6. L Kant, *Prolegommena*, Reclam Biblothek.
7. U. T. I. "The Unification Thought Quarterly," No.7, 1984.
8. W. Kneal, *The Development of Logic*, Oxford. 1978
9. W. J. Einfuhrung in die phanomenologie,(tr, Lee, Yung ho), 1959.
10. So, Hong yul, *Logic and thinking*, Iwha book co, 1984.
11. Son, Bong ho, *Kant philosophy from the metaphysical viewpoint*, Study of Philosophy co,16.
12. U.T.I. *Essence Of Unification Thought,* 1975.
13. U.T.I. *Textbook of Unification Thought,*1983.
14. Ed, Korean phenomenology, Selected Essays in Phenomenology I, Surkwangsa, 1979.

The Presuppositions of Modern Science and Unification Thought

Keynote Speech at the 12th International Symposium on Unification Thought

December 9, 2000, Chien Tan Overseas Youth Activities Center, Taipei, Taiwan

I am deeply honored and highly delighted to be able to speak in the 12th International Symposium on the Unification Thought, which is being held at Chien Tan Overseas Youth Activity Center, Taipei, Taiwan. Today, the topic for the Keynote Speech is: "The Presuppositions of Modern Science and Unification Thought."

I.   The Strategy of Modern Science

We can now address the issue of presuppositions more directly. Bertrand Russell confessed his dismay when his brother explained that in Euclidian geometry there were axioms which had to be taken for granted. We cannot help using certain terms which have definite implications. The important point is to be fully aware of one's assumptions and starting points. I want to note first, Newtonian Mechanism as a presupposition, which has a strong influence on modern science.

Newton thought the main feature of natural philosophy was "to argue from phenomena without feigning Hypotheses, and to deduce Causes from effects, till we come to the very quiet cause, which is certainly not mechanical.…..''

This remark shows his impatience with hypotheses which outrun facts, and his conviction that God is the quiet non-mechanical Cause of phenomena who is immanent in nature. His discipline has proven tremendously influential, underlying as it does modern reductionism and appeals to Occam's razor.

In Newtonian physics, Darwinian biology and Freudian psychoanalyses we may see a common thread of causality. Everything must be analyzed and its cause investigated. Some scientists' success in explanation has emerged from this determined and principled stand. Newtonian physics unified most of the external universes under the general principle of physical science. It was remarkable that gravitation could unite falling apples, the orbiting of the moon around the earth, the orbiting of the planets around the sun, and even the tides in the oceans. The multitude of species and their biological strategies for survival through fitness, duplication and reproduction found an explanation in the theory of evolution.

With the increasing success of science and science-based technology conflicts arose between organized religion and science. With Copernicus, who dethroned the earth as the Center of the Cosmos, Galileo Galilei was persecuted by the Church. Science seemed to give no room for a creator-god to directly intervene in the functioning of the universe. Science's headstrong style is illustrated by the story of Laplace developing a theory of the cosmos in which God has no place. The trend has continued and late twentieth century science has combined Laplace's ambition and Darwin's ideas to produce a picture of an evolving universe endowed with no purpose and developing without any specific divine intervention. Man is not necessarily the crown of creation, but just one of the products of evolution.

I want to note second, the genesis of materialistic reductionism under Newton's Rules of Reasoning, and the preference of science for natural explanations which exclude the 'supernatural.' Even if the physical universe does require a different kind of theory, so-called scientific explanations have become the materialist viewpoint that matter is thought to be the basis of all phenomena, including consciousness. In cases of doubt, the dogma of Occam's razor is invoked.

In practice the razor is used to excise inconvenient mental data, and has a built-in bias towards materialistic reductionism, itself the reducing of the complex to the simple. The important point to note is that Occam's strategy implicitly supports a meta-scientific discipline, just like the proposition that all truth is scientific. Since the 17th century there has been a progressive merging of the spiritual dimension into the physical, a reversal of the traditional properties of being.

I want to note third, anti-teleological determinism. We saw Galileo abandon the teleological approach for efficient causality, neglecting the 'why' for the 'how.' Thus, anti-teleological determinism refers to the triumph of the mechanistic view over the animistic, of matter over spirit. Newtonian physics and astronomy laid down mechanical laws, and, though the rigor of the determinism has been somewhat weakened by quantum indeterminacy, still mind is regarded as dependent on the brain. Very few neurologists have dared take a mentalist or dualist position, which does justice to the intentionality of consciousness or mind.

Finally, I want to note the influence of positivism. Logical positivism has been discredited by the Popperian replacement of verifiability by falsifiability as a demarcation between science and non-science. In explaining his criterion of falsifiability, Popper distinguished scientific and non-scientific or metaphysical theories. Yet logical positivism still remains powerfully influential on ways of thinking. Positivism came about as a reaction against theology and metaphysics. It laid down verification principles which prohibit seeking for ultimate causes beyond the material realm which is the domain of natural science. Scientific investigation is seen as the triumphal march of materialism toward a complete explanation of life.

II.   The Approach of Unification Thought towards science

Scientific knowledge is essentially objective knowledge, which has been perceived to eliminate the subjective element of observation. Therefore scientific knowledge is a knowledge derived from the reasoning mind of man. Value judgment, metaphysical theory and aesthetic consideration may be merely personal, but they may also be supra-personal. Religion, metaphysics, ethics, and the arts are objective in the sense that they seek both to understand and know about the world. Scientific theories are in principle testable and therefore reasonable, while metaphysical theory can be repeatedly redefined and criticized. Spiritual knowledge is knowledge about the spirit as object which can be gained by the reasoning mind of man.

In the Unificationist view, science and religion have a common share in the pursuit of truth. The object of scientific pursuit is external truth in order to move mankind from outer ignorance to outer knowledge, while religion searches for internal truth to move from inner ignorance to inner knowledge.

Scientific knowledge is characteristic of pure knowledge-orientation, whereas Unification epistemology is characteristic of knowledge-orientation together with a purpose-driven end (telos). Paradigmatically, though western scientific epistemologies are fact laden, the Unificationist one is fact-value. It may be correct to say that Unification epistemology is characteristic of knowledge-oriented teleology.

Generally speaking one could say, 'science is value-neutral,' but this value-neutrality of science is itself a value-judgment. Maxwell equates such value-neutrality with value-blindness, or value-insensitivity. In western philosophy in general, any conflation of fact and value brings about a confusion of categories and value assertions bring about empty-talk in the sense of logical positivism.

But, because of the conceptual model related to value orientation in Unification Thought, this category mistake is avoided because "value and fact cannot be separated." Two different paradigms (the western and the oriental) are implicitly contained in Unification Thought. That is, a non-contradictory possible fusion of fact-value is the case.

Against the positivist background, we might quote the following from Essentials of Unification Thought: "Absolute truth refers to the universal eternal truth. Therefore, without having the Absolute Being as the standard, the concept of absolute cannot be established."

The word of absolute truth in Unification Thought context cannot be verifiable in the sense of positivism. Contrary to positivism, absolute truth in the Unification context contains two aspects of truth, the one is value-laden truth with a kind of supermundaneness, the other is fact-laden truth (knowledge) with a kind of mundaneness. Thus, value and fact cannot be separated even in this case.

We reject the presuppositions of modern science, which lead to mechanism, reductionism and materialism, not because they threaten religion but because they are fallacious strategies, which by demolishing all metaphysics, demolishes the very science they set out to account for. Both science and religion are human enterprises, which pursue truth, arising from puzzlement about this world. In the light of Unification Thought, we can see a prospect of harmonizing science and religion.

It is my hope that through this symposium, Unification Thought will be developed into a movement of science. The centripetal point for shaping science is Unification Thought, and we need to tap the possibility of unifying the various thoughts with Unification Thought as the axis.

Thank you very much!

PURSUING A NEW MODEL OF SCIENCE

KEYNOTE SPEECH AT UNIFICATION THOUGHT SESSION, WCSF 2001

JANUARY 29, 2001, NEW YORK HILTON, NEW YORK, USA

I am deeply honored and highly delighted to be able to speak at this Convocation of World Leaders, World Culture and Sports Festival 2001, Unification Thought Session, here in New York City. Today, the topic is: "Pursuing a New Model of Science," in relation to Unification Thought. I can now address the issue of the presuppositions of modern science, which have certain definite implications. We cannot help using certain terms which have been taken for granted, such as, the axioms of Euclidean geometry. There are four of these presuppositions that I wish to note: First is Newtonian Mechanism, which has had a strong influence on modern science. Second is the genesis of materialistic reductionism under Newton's rule of Reasoning, and the preference of science for natural explanations which exclude the supernatural. Third is anti-teleological determinism, which refers to the triumph of the mechanistic view over the animistic, or of matter over spirit.

Finally, I want to note the influence of positivism, which came about as a reaction against theology and metaphysics, and laid down the verification principle which prohibits seeking ultimate causes beyond the domain of natural science. Scientific investigation is seen as the triumphal march of materialism toward a complete explanation of life. In contrast to these hitherto dominant features of science, I want to point out some new trends in modern science, in order to pursue a new model of science based on Unification Thought. One thing becoming clear is that modern science has begun to take a point of view of vitalistic holism, which goes beyond the dichotomies in the thought of Descartes or Newton. That is, modern science is beginning to go beyond the limitations of a mechanistic view of the world, in which subjectivity and objectivity, material and spirit are separated, and beyond the limitations of atomistic reductionism. Prigogine, a winner of the Nobel Prize in Chemistry in 1977, asserts that modern science must be newly established based on the law of the increase of entropy, the second law of thermodynamics, transcending the concept of reductionism that means precise measurement based on the theory of causality from classical physics. He says that people must retrieve the Greek method of thinking that described the cosmos as a work of art, and escape from the classical method that describes it as an automatic mechanism.

David Bohm, a quantum physicist, and one of the leading scientists of the movement of new science today said there has been a realistic, atomic or mechanistic view of the world that thought of material and spirit separately. However nobody will deny that since the emergence of quantum mechanics there must be an implicit order behind matter and spirit, related to oriental thought." He thought that matter and spirit are not two separated beings but one substance of dual aspects on the higher level in the world of implicit order. Although knowing computer, information and cybernetics theories is the prior condition to understanding information today, it is information that takes the common central role among those three theories. Communication through this information is definitely needed to realize some purpose. Until now science has been based on mechanistic methodology excluding any teleology. However, modern information science has introduced the concept of finality in science. Recently cerebral physiology has emerged as the newest science. According to this new trend, the concepts of purpose, symbol, cerebrum, central nervous system, language, etc. has become the main concern of modern science and philosophy, all relating deeply to the human spirit. Heisenberg thought that one vibration of a particle can influence the whole cosmos. His uncertainty principle stated the magnificent fact that the atomic nucleus is closer to the vibration (rhythm) that occurs relating to the entirety rather than merely to the material. Quantum was nothing but a wave in the quantum field. Koestler asserted the revolutionary theory of the holon which was similar to the monadology of Leipniz. The monad of Leipniz has a structure completely different from the continuum of space and material. It is the entirety as a unified organization involving the whole internal being of the individual as the material and spiritual being in itself. The Holon is a mini-entirety, that is, it is not a part separated from the whole but the part between the whole as well as the whole within the part. In other words, we cannot separate any object from itself but neither from the rest of the Universe: any object is at the same time the one and the whole. This is true for any object in the universe, as well for an elementary particle as for a complete organism. There exists a certain aspect of the world that cannot be explained by the mechanism of causation. The content of the experience of modern physicists is beyond the realm of the classical concept of space and time. This experience can only be understood by new concepts of space, time, and matter. According to Unification Thought, God's own dynamic unity of Sung Sang and Hyung Sang centered upon His purpose generates acting energy to unite all things in the world, i.e., "to make all things interact with one another." Let us now note the Unification Thought view:

1.   The Unity of Matter and Mind,
2.   The Unity of All Things,
3.   Value Oriented Scientific Knowledge.

The first two points are similar to the point mentioned earlier concerning the new trends in modern science. Finally, I want to raise the third point. Scientific knowledge is characteristic of a pure knowledge-orientation, whereas Unification epistemology is characteristic of knowledge-orientation together with a purpose-driven end (telos). Paradigmatically, while scientific epistemologies are fact-laden, the Unificationist one is fact-value-laden. It may be correct to say that Unification epistemology is characteristic of value oriented scientific knowledge. Generally speaking one could say, science is value-neutral, but this value-neutrality of science is itself a value judgment. Maxwell equates such value-neutrality with value-blindness, or value-insensitivity.

In western philosophy in general, any conflation of fact and value is thought to bring about a confusion of categories, and value assertions themselves are thought to bring about empty-talk in the sense of logical positivism. But, because of the conceptual model of value orientation in Unification thought, this category mistake is avoided because value and fact cannot be separated." Two different paradigms are implicitly contained in Unification Thought. That is, a non-contradictory possible fusion of fact-value is the case. For the purpose of pursuing a new model of science, we must go a step further and see that the relationship of matter and mind, the one and the whole, and fact and value, cannot be conceived as a state of virtually exclusive opposition, but as two aspects of the same reality, which co-exist and are in continual co-operation. May your discussions at this Unification Thought Session be very fruitful. I pray for God's protection and blessing to be with you and your families. Thank you very much.

Materialism, Dualist-Interactionism, and Unification Thought

Keynote Speech at the 13th International Symposium on Unification Thought

December 1, 2001, Hotel Praha, Prague, Czech Republic

I. A critical survey of materialism

In recent centuries science has had astonishing success in providing explanations of a wide range of natural phenomena. The appeal of the physicalist or materialist philosophy of the brain-mind problem lies in the claim that it is based on natural science. It would seem foolish and reactionary to resist the physicalist claim that eventually all phenomena, including even the most subtle of human thoughts and actions, will be completely explained scientifically.

Ever since Darwin, it has generally been believed that the evolution of human beings is by natural selection, an entirely materialist process. The story of a divine creation was alleged to be scientifically untenable. Materialists state that in evolution, consciousness emerged when brains became sufficiently complex, and that with the further evolution of humanoid brains, and speech, there emerged a higher level of consciousness, namely, the self-consciousness. It was believed that evolutionary development produced a human brain that could deliver all of the higher spiritual performances of creative art, literature, philosophy, religion, and creative science. This idea gained general acceptance because it conformed with pervasive materialism and with Darwinian evolution.

However, this belief was never adequately formulated or critically examined. According to materialist theories of the mind, mental activity is an attribute of matter, or of the physical world; and of all matter of higher state. The activity exists in the organized nervous systems of higher animals and man.

Typical sorts of materialism include:

Negative materialism and behaviorism, arguing that there is no spirit, as developed independently by Skinner and Quine.

Reductionistic materialism in which mind is only an activity of materials, as developed by Hobbs and Feuerabend.

Creative materialism arguing that spirit is an advanced creative activity arising from the action of the brain, as advocated by Diderot and Darwin.

The materialist account is usually some version of the identity theory by Feigl.

According to identity theory some neural activities in higher levels of the brain, the cerebral cortex, have linked mental or mind-like states. These are alleged to give us all the richness of our conscious experiences. This is merely dogmatism made to account for mental experiences as merely derivative from human neural activity, in an effort to save materialism.

Great display is made by all varieties of materialists that their brain-mind theory is in accord with natural law as it now is. However, this claim is invalidated by two most weighty considerations. First, nowhere in the laws of physics or in the laws of derivative sciences, chemistry and biology, is there any reference to consciousness or mind. Regardless of the complexity of electrical, chemical or biological machinery, there is no statement in the 'natural laws' to account for the emergence of this strange non-material entity, consciousness or mind. Second, all materialist or physicalist theories of the mind are in conflict with biological evolution.

Since all materialism (radical materialism, epiphenomenalism, and the identity theory) asserts the causal ineffectiveness of consciousness per se, they fail completely to account for the undeniable; biological evolution of consciousness. Materialists have no explanation for the origin of self-consciousness, and do not recognize this failure.

The mind-body problem leads us to the mental intentional disposition in the field of philosophy and science. As an example, in philosophy Brentano characterized the essence of self-report to be the ability to distinguish between the perceiver and the perceived. This principle is usually referred to as "intentionality." In physics Heisenberg has clearly stated that the science of physics, dealing primarily with both the observer and the observed, must be inferred. Gilbert Ryle first defined mind in terms of minding, a function. Minding is the behavior of paying attention; and it is a considerably important subject of scientific knowledge concerning behavior and attention.

Materialism usually ignores those intentional aspects of the mind-body problem.

The mind-body problem is fundamentally caused by ontological reasons of their own. Traditional ontologies are centered with reason/spirit, or with will, or with material matter itself. Moreover some of them claim the monistic assertion that spirit alone is substantial or that material matter alone is substantial.

According to the ontology of Unification Thought, the core attribute of all beings is heart. Sungsang and Hyungsang is only one aspect of entity of all beings. Centered on heart (purpose), Sungsang and Hyungsang engage in give-and-receive action. That is the way all beings exist.

Same as the ontological ground, cognition and practice can never be made based solely on the physiological process of the brain. this is because cognitive action takes place through the give and receive action between the mind and brain.

II. Eccles' Interactionism in view of Unification Thought

Dual entity properties of the mind and brain

Unification Thought asserts, based on its theory of the dual characteristics of the Original Image, that all beings have dual characteristics, namely, Sungsang and Hyungsang. The human is a dual being of mind and body, and the cells, tissues, and organs making up this human body are composed of mental and physical elements as well. Furthermore, all human actions and operations are dual, which means that psychological and physiological actions are always at work in parallel.

Therefore, from the perspective of Unification Thought, psychological and physiological processes are always at work in parallel during cognition. This means that mental activity occurs through the give-and-receive action between the mind and the brain. In contrast to materialist or physicalist theories of the brain-mind problem the alternative theory of dualist-interactionism gives opportunity for the conscious self to exercise free choice between genuinely alternative modes of action, or to achieve the expression of some creative insight. The essential feature of dualist-interactionism is that the brain and the mind are independent entities, the brain being in the matter-energy world and the mind in the world of conscious experiences.

Eccles advocates the theory of the interaction of the mind and body by making a distinction between the brain and the self-conscious mind. The contents mentioned below are those of his book, Self and Its Brain, in collaboration with Popper.

1. Each experiment in a self conscious mind has unique characteristics in itself so that we can say that any movement in the brain proper places the focus on any activities in that particular moment and space. This fixing focus, as it is called, is attention.

2. The experiment of the self conscious mind is assumed to have something to do with neural phenomenon of the liaison brain. Nevertheless, it doesn't manifest the identification, but performs its reciprocal action indicating its correspondence to some extent.

3. There is a laps of time between the experiment of the self conscious mind and neural phenomena.

4. A continuous experiment is in existence where the self-conscious mind works upon the brain proper.

To sum up, Eccles opposed the identity theory which identifies neural activities with the thinking process itself. He sees evidence of the reciprocal action of the brain with the mind in the conscious awakening of the brain proper in such things as voluntary movement, decoding of any verbal expression, or trying to manifest ideas.

Mind-Brain interaction based on quantum dynamics

As we have seen, the closed system of the physical world has been safeguarded with great jealousy in all materialist theories of the mind. The closed system of the physical world is a key feature of all materialist and physicalist theories of the mind, for the mind is postulated to be an integral attribute of special parts of physical world, namely the brain.

Necessarily the dualist-interactionist theory is also in conflict with present natural laws, that is conservation laws. Thus there must exist a frontier between the physical world and the mental world, and across this frontier there is interaction in both directions, which can be conceived as a flow of information not of energy. Thus, we have the extraordinary doctrine that the world of matter/energy is not completely sealed, which is a fundamental tenet of classical physics, the conservation laws.

Still, there is the question how it is possible for a self-conscious mind and a brain to connect. Eccles at first made a patient with separated brains an object of his scientific study, through which he discovered that the left half hemisphere does the work to do the liaison job. He found that only the left half of the brain has the ability to speak a language and, has conceptual ability. (Since then, the right half has also been found to have such an ability, regardless of the differences in qualification.)

Eccles then revealed that only a very tiny area in the realm of the cerebral cortex directly effect liaison (brain) with the self conscious mind. That is, Eccles thinks that the intermediary part is a functional conically structured module about 3 mm long and 0.1 mm to 0.5mm in diameter rising vertically from the surface of the cerebral cortex.

Moreover, the interaction activity, across the frontier, need not be in conflict with the first law of Thermodynamics. The flow of information into the modules could be effected by a balanced increase and decrease of energy at different but adjacent microsites, so that there was no energy change in the brain.

Eccles accepted the views of H. Margenau's book, The Matter of Mind and Brain, professor emeritus at Yale University in the field of physical philosophy and physics. Margenau approached the problem of the reciprocal action of the mind and brain inspired by thought, in the field of probability in quantum dynamics, where there is no energy or physical matter. Therefore, in explaining the reciprocal activity between mind and brain we don't need any energy to be put in. Following Margenau, the hypothesis is that mind-brain interaction is analogous to a probability field of quantum mechanics, which has neither mass nor energy, yet can cause effective action on microsites.

In view of Unification Thought concerning mind-brain interaction, I'd like to define life at first. Life is cosmic consciousness that is injected into cells or tissues. When God created the Universe through Logos, He inscribed all the information peculiar to each living being in the cells of that being as a material form of code. When cosmic consciousness enters into a cell, it reads the genetic code of the DNA of the cell. More over, in the human body information from each of the cells and tissues is transmitted to the center through the peripheral nerves; and order (information) is transmitted from the center through peripheral nerves to the cells. Like an exchange of information inside our body, mental consciousness (mind) and brain interacts reciprocally through information. This pattern of explanation of Unification Thought is akin to the Eccles' interactionism. Through the speculation of the Eccles' interactionism, cognitive process centering on mind-brain interaction in Unification Thought will be proof of an authentic theory of science.

The Mind, Spirit World and I

Keynote Speech at Unification Thought Session, WCSF 2002

February 15, 2002, Hilton Hotel, Seoul, Korea

Ladies and gentlemen, I am deeply honored and highly delighted to be able to speak in the International Symposium on Unification Thought. I want to thank you all for your participation in this symposium. The advance of Unification Thought and the growth of a world level culture wherein all people are capable of living together in harmony and creativity depend on the efforts of all of you. The seminal work of Rev. Sun Myung Moon and Dr. Sang Hun Lee form the basis of our work, but it is your effort and thought, blessed by the grace of God, which will bring the ideal into reality. Please accept my humble remarks as part of this larger attempt to bring the ideal into being.

I.   Mind / Body Problem As Seen By Science

Looking back on the history of philosophy and science, the issue of the contrastive duality of mind and body remains one of the hard questions for all ages and cultures. It is, however, plainly true that the general tendency of the current new science is to take the standpoint of a vitalist totalitarian theory after having reached the top of the mechanistic world view and atomic reductionism, which separated body and mind into material and soul in the dichotomy taught by Descartes. Heizenberg's discovery of the principle of uncertainty marked a turning point in history in its recognition that the atomic nucleus is not matter but a vibration occurring in relationship with the whole. A quantum is only a quiver on a quantum field. He noted that Bohr's theory of the interdependencies within wave-particle mechanics is very similar to Descartes' notion of the duality of mind and body, and further noted that even though there is spiritual nature among the contents of cognition, it is deeply relevant to the nature of the brain, which is physical. David Bohm, a quantum physicist and advocate of present day new science, expressed the idea that matter was thought to be different from the spirit in the mechanistic world view as well as in realist atomic theory after Newton.

Since the development of quantum dynamics, however, it is known that there is implicit order behind both matter and spirit. He said that matter and spirit are not two in the world of the implicit order, but are dual aspects of a reality in a higher dimension. A key element of all materialist and physicality theories of the mind is the closed nature of the physical World, for the mind is postulated to be an integral attribute of certain special parts of the physical World, namely brains. On the contrary, contemporary biology of mind/brain interactionism has put forth a biological field theory, which has borrowed from the field theory of current physics. Seen from this theory, there is a frontier (brain liaison) between the brain and the mind, and across this frontier there is interaction in both directions, which can be conceived as a flow of information, not of energy. As stated by Popper (Popper and Eccles, 1997, Dialogue XII), interaction across this frontier need not be in conflict with the first law of Thermodynamics, which is a fundamental tenet of physics.

II.   Concerning the Problem of The "Self"

One of the central problems in contemporary theoretical physics is the interaction between an observed phenomenon and the consciousness of the observer considering the phenomenon. We could say that physics has accepted a clear distinction between the subject and the object, between mind and matter. But the mind is an object itself, in the view of Jean E. Charon. To be sure, thought (or mind) is not directly observable like an ordinary object, but it is representable. Let us agree to call observable phenomena "reality" (extension), and non-observable phenomena "the imaginary" (thought). Thus, at the cosmic level, black holes, which represent the final evolutionary state of a great number of stars, are non-observable. Such is also the case, at the microscopic level, with elementary particles or quarks, which enter the constitution of nucleons composing the nuclei of any atom of our world. These objects can truly qualify as imaginary objects. The discovery of the imaginary by physics can truly be compared to the discovery of a new world, indeed the rotation from a material to a spiritual world. The subject, the "I," was situated at the "intersection" of thought and extension. The "I", the subject, was not thought, but the one able "to look at" objects, that is, to look at both thought and extension. Popper conjectures that the self is not a pure ego, that is, a mere subject. What's more, it is incredibly rich. It is acting and suffering, recalling the past and planning and programming the future. It contains, in quick succession or all at once, wishes, plans, hopes, decisions to act, and a vivid consciousness of being an acting self, a center of action. The soul is regarded by Christian tradition as the spiritual part of man.... it is bound up with the essential personality of each individual, and each soul is unique; it is also responsible for moral decisions and rational conclusions, and is immortal. This divinely created psyche should be central to all considerations of immortality and of self-recognition. In all materialist theories of the mind there can be no consciousness of any kind after brain death. Immortality is a non-problem. Presumably this arises from the pervasive materialism that is blind to the fundamental problems arising in spiritual experience. But this whole cosmos is not just running about with no meaning.

III. The Spiritual World as written by Dr. Lee

I want to introduce a summary of the spiritual world written by Dr. Lee, which was not completed while he was alive as follows: It is the most important factor in realizing the value of our lives to know the spirit world—which is the substantial subject world to the physical object world—and live according to the principled relationships of the world. The cosmos was created patterned on the dual characteristics of God, and the spiritual or heavenly world is the incorporeal world corresponding to the human mind, while the physical world is the corporeal substantial world corresponding to the human body. The physical world can be felt and perceived with our physical senses and physical reason—physical senses and reason feel and perceive practical phenomena. In the same way, the spirit world can be felt and perceived with the spiritual senses and reason. Together these two aspects of any human being make up the "microcosm."

The physical world has the purpose of serving as a training ground for the life in the spirit world. It is in the physical world that we accomplish the three great blessings—conjugal love centered on God, human love centered on God, and the love of nature centered on God—and learn these lessons of love implicit in the four-position foundation, without which life in the spiritual Kingdom of Heaven is impossible. The purpose of this spiritual world is an eternal life of love lived in attendance on God (True Parents) and the life of the four ways of heart centered on the four-position foundation, that is, life should be happiness in itself. The world we live in today, however, is the unprincipled fallen world, and people's lives can be distinguished according to the value of their love. We can distinguish four categories according to this value.

1. The Lowest Order, Having No Value: This class consists of those who live for their own benefit, e.g., swindlers, robbers, thieves, murderers, rapists, etc. These are those who spent their whole lives in pursuit of wealth or sex without any value. 2. The Second Order, With Lowest or Ba-

sic Value: These do not know or believe in God, but still follow their con-science. Though often doing wrong, they still try to do the right thing. 3. The Third Order, With Middle Level Value: These believe in God, try to perceive His Will, and try their best to realize goodness. 4. The Fourth Or-der, With Highest Value: These know, not only the being of God and His will, but also His love, and teach them to all people through their own liv-ing example. This order is exemplified by many of the founders of relig-ions.

There is, in addition a special class consisting of those blessed families who have accomplished the four-position foundation centering on God. There are four divisions in the unprincipled spirit world, which correspond to these orders of value: a middle or antechamber level, hell itself, a form spirit level, and paradise (or life spirit level). These four do not necessarily correspond exactly to those in the physical world. 1) Hell, corresponding to the lowest order of value is below the middle spirit level. Those who lived self-centered lives in the world live there. Hell is further divided into three levels. It is gloomy, smells bad and wet, and all there is in a wretched plight. The record and length of one's earthly life without value determines where one stays in hell.

2) The middle or antechamber level, corresponding to the second order of value is where newcomers arrive after death. It is right above hell. Here spirits evaluate their lives and decide where in spiritual world they belong. All the secrets of a person's life are revealed here. All habits from the physical world disappear over time and spiritual law begins to be a per-son's guide. Here, no kind of worldly honor, knowledge, power, desire, position, or wealth are of any use. On the contrary, truth, beauty and good-ness centering on true love become the basis of valuation.

3) Formation stage spirit world, corresponding to the third order of value, is right above the middle spirit world. Here go those who belonged to the second class, such as men of conscience, charitable persons, patriots, teachers, men of virtue. According to Divine Principle, these are form spir-its who accomplished the formation stage level of character. The spirits in this class correspond to the righteous from the Old Testament age.

4) Paradise, corresponding to the highest order of value is the next highest level and those who belonged to the third order go there directly through the middle spirit world. Paradise is also divided into three parts: low, middle, and high.

Those four divisions constitute the unprincipled spiritual realms. Prop-erly speaking the only portion of the spiritual world which is fully princi-pled is the Kingdom of Heaven itself, and it is here where those blessed families who have accomplished the four-position foundation centered on the Messiah go. Conclusion concerning on spiritual world; 1) the spirit world exists practically and is the world of principle and order. 2) The spirit world is a partial spirit space of the cosmos created by God, who is absolute, unique, unchangeable, and eternal. It has existed, exists, and will exist as a spiritual environment for the object of the beauty and love of God, and is the place where the spirits of human beings can return pleasure to eternal God with the fruit of their accomplishment in the earthly world.

Furthermore, that world is the substantial subject world to the physical object world. In short, the spirit world is the vertical causal world and the physical world is the horizontal resultant world. Accordingly, the spirit world continually influences the physical world and is influenced by it. The spirit world and the physical world exist in a give and take relationship. No development or change in the physical world can exist without relationship to the spirit world. Any final result in the physical world will be revealed and preserved forever in the spirit world. Now, all of us in the world should awaken to the reality of the spirit world and accelerate our efforts to reach true love and true value. We should live a life for others through sacrifice and service until we die. Embodying true love through doing goodness and training ourselves is the first priority for all of us in this world. Thank you very much.

Unity of Academic Disciplines and Unification Thought

Keynote Speech at Unification Thought Symposium, WCSF 2003

July 10, 2003, Sun Moon University, Asan, Korea

Most honorable and distinguished scholars, respected guests of this symposium on Unification Thought, Ladies and Gentlemen: It is a distinct honor for me to address you the topic Unity of Academic Disciplines and Unification Thought at the beginning of this very significant international symposium. In order to be here, you have taken time from your work and important schedules. I know this indicates your awareness that our world is faced with many serious challenges, and that it is our responsibility to work together in a search for solution to the critical problems we face today.

I. Two Types of Unity of Science

A) The Type of Causation—Reductionism

Through the movement of unity of science, we know that scientists have tried to carry out the task and ideal of science for a long time. To sum up the history of that movement from the Unification Thought view, the first attempt tried to realize the ideal of Unity of Science along with the developing science, which was the movement of the School of Encyclopedia based on the philosophy of enlightenment in the seventeenth century. They compiled an encyclopedia on the basis of empiricism of the philosophy of enlightenment and mechanical materialism, which is the predecessor of The Encyclopedia Britannica enhancing its reputation even today. The second attempt was the movement of Unity of Science centering on the Vienna Circle. The academic trend in the School of Encyclopedia or the Vienna Circle was originated by the tradition of empirical epistemology and materialistic philosophy, and the two perspectives were strong advocates of empirical reductionism. I will call these traditions "the Type of Upward Causation."

This type of upward causation had a tendency to base epistemological foundation of all sciences on empirical sense data and to base ontological foundation of all sciences on materialism. The atomists were the first reductionists. Democritus' atomism, certainly very popular in his day, was plausible, not because of the mechanistic postulation of atomic discontinuities, but because of the assumption that a very small stock of accessible concepts like shape, size, and speed could in principle explain all change and all diversity in the natural world. It assumed that the most complex properties of life and mind could, in principle, be understood in simple quantitative terms. The success of Galilean mechanics was the second attempt at reductionist materialism. In his view, secondary qualities (like color and viscosity) could be explained in terms of primary qualities (like extension and mobility, which warranted the application of mathematized mechanics). Thus the language of science would constitute no problem since it would ultimately reduce to a set of quantitative empirical concepts. In this view, the unity of science must be attainable because all sciences ultimately reduce to the most general science, physics (physicalism).

Such concepts would constitute the universal principle of the unity of science in the Vienna Circle. By making science a single enterprise, epistemologically speaking, the unity of science could be achieved. The original thesis of the unity of science was based on the belief that all scientific concepts are definable in the strictest sense of language of sense data by terms belonging exclusively to the positivistic language or, alternatively, the physical language. The question whether intentional and other non-extensional logics are reducible to extensional ones to which were for early Wittgenstein and Russell a matter of deep logical conviction, which was dealt concretely with by the Carnap. He originally formulated the thesis of extension-ability as one of the mainstays of the unity of science in a very calm and definite way. Today these claims of reducibility look rather naive. Most logical empiricists came to realize that most terms occurring in scientific theories are neither definable by, nor reducible to, the so-called observational terms of the everyday thing-language. The Institute for the unity of science has recently dissolved. That its aims have been achieved might well be doubted. The original positivist manifesto has been abandoned. Though many would still set up a methodological unity of science as a goal, the barriers to the unification of the natural and the social sciences are serious ones. These are matters of the most intense controversy among philosophers of science at present.

B) The Type of Downward Causation—Unificationism

The two movements for the unity of science through the development of science, above mentioned, did not play a leading role in realizing the unity of science. If I point out the causes of the failure of the two move-ments from the standpoint of Unification Thought, we could say that they were founded on a mechanistic world-view, or the premise of the positive philosophy, and that they advocated that premise dogmatically and intoler-antly. The unification of science is a work that can never be attained from a dogmatic or intolerant initiative. Accordingly, the unification theory of sci-ence can be actualized when it is based on a more comprehensive and uni-versal premise and philosophical foundation. It is for this reason that we need this kind of philosophy. If science is viewed as an ensemble of theo-ries, the problem of the unity of science has traditionally been dominated by the idea of reduction, that is, upward causation, which means that theo-ries about a higher ontological level can be reduced to a theory about a lower ontological level, which is the basis of the final cause.

In short, higher and more comprehensive scientific theory in the sphere of ontology should be reduced to lower and more basic scientific theory in the sphere of ontology. It is due to the belief that the language and scien-tific theory of the sensory data of physics, the lowest ontological dimen-sion, is the fundamental cause of all studies. However, contrary to reduc-tionism, downward causation is also important. As reductionism, upward causation is persuasive when an atomist explains a mechanistic world-view such as materialism and behaviorism, downward causation is also more persuasive in explaining an organic world-view of teleology such as ideal-ism and spiritualism. For example, knowledge about the cellular control of molecular activities may complement knowledge about the molecular con-trol of cellular activities; that is, the higher cellular level controls the lower molecular level. As shown by N. Chomsky, the facts of language acquisi-tion by children are better accounted for by assuming a special innate fac-ulty of language based on mentalism than by relying on behaviorist learn-ing theory. Theory of N. Chomsky's linguistics presents a model of the typical downward causation through removing the language ability of man from the innate faculty his spirit obtained.

From this viewpoint, Unification Thought is also considered a down-ward causation centering on God and spiritual value. In Unification Thought, the mind and body of a man is in unity, naming the former and the latter as subject and object, cause and effect, and vertical and horizontal relationship respectively. In this case, mind-body is a causal cause-effect, so the causal cause is established by the subjective mind. Therefore, the subject-object relationship becomes a model of downward human relation-ship having the subject as a cause. For example, in a human body, the sub-consciousness of life (mind) is the subject, and the organization and the cell of a human body becomes the object. Judging from the law of cause and effect, the subjective life (mind) becomes the cause to control and dominate the objective body. This subject-object relation model is not lim-ited to the relationship of mind and body but becomes the most fundamen-tal principle to control the existing world of man and nature. To say it once more, all beings existing in the world are a methodical organic body connected by causational chains of subject and object.

The human world as well as the natural world is composed of a grand methodical unified body. This principle to maintain the methodical system of subject-object in nature is natural law, and the one that maintains the methodical system of subject and object in human relationships is the law of value or ethical law. Therefore, it is the task of all studies in science, philosophy, and arts and so on to research the laws controlling these two worlds. In this case, the subject holds priority over the object in the sphere of ontology and axiology. It is because the subject holds the subjective, affirmative, centripetal, and vertical value, that the object holds superficial, passive, peripheral, and horizontal value, to maintain the reciprocal relation. As there is no body without mind, there is no science without value. This is because the relationship between value and fact is the necessary relationship of the subject and object. It can be seen that the ideal of unity of science, which failed to actualize in the way of Reductionism (Upward Causation), can be actualized in the way of Downward Causation. It is, therefore, said that Unification Thought is a theoretical system that restores a real position of academic disciplines and truly unifies all academic disciplines by properly establishing the relationship of subject-and-object in the law of cause-and-effect.

II.   Unification Logical View in the Unity of Science

Aristotle's attempt is the first to construct a system of all sciences based on formal logic. Science in his view must rest on first principles, themselves necessary truths, certified as true by the skilled insight of the philosopher-scientist. From these as premises, deductive logic can generate demonstrative science, in which the more specific conclusions appear as necessary truth. Lacking such principles, there is no way in which an eternal and necessary science could be constructed. Aristotelian logic provided a tight deductive and conceptual unity of science, since all conclusions have to be implicit within the premises with which the deductive inference begins. Particularly, the formal logic of Aristotle has been regarded as self-evident truth. But, is it really so? If logic aims at setting the foundation for a proper scholarship of the sciences, it must be admitted that the foundation of the 'formality' in formal logic is weak. This is because we cannot assert the universality of the sciences from the abstract 'formality' which originated in empirical positivity.

Unification Logic recognizes the a priori (priority) sphere of subjective conditions as the foundation of the possibility of such positive experiences and of those which control all empirical conditions. Accordingly, from the viewpoint of Unification Logic, the foundation for proper sciences can begin neither with empirical 'positivity' alone nor with the subjective a priori alone. It is Unification Logic which is being promoted as an alternative to formal logic, which has been the basis of most academic endeavors. As seen above, Aristotle's deductive logic could not become the universal principle to establish the foundation of metaphysics and science. From the viewpoint of Unification Thought, the system of formal logic of Aristotle is literally formal system, which is a vacant logic system to tell nothing about the empirical essentialities and empirical science, which can become the basis of science.

To establish an eternal and necessary foundation in the view of Unification Thought, as Aristotle mapped out, the deductive motive coming from the mind, sphere of sung sang, and the inductive motive of the level of experience, that is, sphere of hyungsang should be under consideration. In Unification Thought, when these two deductive and inductive motives are under consideration in the sense of subjective and objective, the basis of true science (metaphysics and physics) can be secured. Kant saw no way of explaining the universality and necessity of the law of Newtonian mechanics, other than that he supposed that the source of this universality and necessity lay in the human mind itself. As seen from Hume's inductive argument, no particular sense could of itself achieve the certitude of universality and necessity. The mind constructs in advance, for example, the categories of a 'pure physics' of synthetic a priori truth. Kant was forced to look carefully at the mind's power of projecting its own structures into the objects of scientific inquiry. He believed that the mind is already equipped structurally with the categories of 'pure physics' meaning the synthetic a prior truth. Through that, he thought the universalities and necessities of empirical science could be regulated from the innate a priori structure of mind.

From the viewpoint of Unification Thought, the typical examples of the universal principle that can be a basis of science is the principle of interaction of give and receive. Science is a process of seeking truth to try to explain individual experienced facts through a general and universal principle. Accordingly, the work to establish science cannot be achieved in the sphere of experience or in the sphere of mind either. It is because the dimension of mind and the dimension of experience are matched in the reciprocal relationship of subjectivity and objectivity. So, neither of them can regulate the other. The unique principle to work between both parties is the principle of give-and-receive. If you search for the principle of regulation in mind only, the empirical world is nothing but a chaotic various collection of sense data. If that were so, wouldn't the mechanic law of Newton apply in this real world? Are all those laws subjective mind laws? Do we need a fundamental principle to regulate the experimental physical world? Unification Epistemology maintains that man's subjectivity is involved in cognition. Of course, the sensible qualities of the object, also, are necessary for cognition to take place. Accordingly, the general principle that science seeks for appears in the mind as the law of representation and practical law in the world of experience, which are regarded to have a relationship of subjectivity and objectivity. Those two subjective and objective categories include the relationship of subjectivity and objectivity, and the principle to regulate the spiritual world of mind and physical world of experience respectfully. From this viewpoint our cognition can obtain the ability to recognize the physical laws of the empirical world, and accordingly, embrace all the subjectivity and objectivity simultaneously. In this respect, Unification Thought stands above Kant. Furthermore, it is possible to establish the theory for the unity of sciences on the structure of the logical foundation of Unification Thought.

III. Common Ground of Religion and Science

Central notion on the positivist theory of science is the concept of verification. As you know, positivists' presumptions have had a deep and lasting influence on the general intellectual climate of our time. We have seen that the verification principle as an extension of this thesis became an arbiter between sense and nonsense and therefore an essential criterion of rationality. Theological assertions, because unverifiable, are thus rendered nonsensical and therefore irrational. Consequently, theologians who opposed this verification in science, and its extension as a general principle of rationality, felt uneasy about the problem of verification in religion. But the theory of science is not the process of an inductive generalization to collect data and evidences through inductive work, but is a scientific thinking process to present hypotheses through a bold conjecture, intuition and imagination.

Rather, science is the discipline which intensifies:

1. Imagination,
2. Critical discussion, and
3. Experimental testing in the quest for truth.

The point is that, without daring and bold conjectures and hypotheses that grow out of the attempt to solve cognitive problems, we could not improve our knowledge. Religion and science do not differ with regard to vision and imagination, for either will perish without vision. Belief is also a major part of the scientific process. If science were nothing but skepticism, nothing would ever come of it, that is, no scientific advances would be made. Belief is the tendency to act on the basis of expectations.

When scientists test their hypotheses and conjectures on those expectations, they are then living up to the scientific ideal. So far, religious commitment and scientific commitment are seen to be psychologically very similar. Both require belief. But this leads to the question of the role of doubt in the form of severe critical discussion and crucial experimentation.

It is sometimes said that science differs from religion in that science requires critical debate, whereas religion tend to resist it. But this is oversimplification. It would appear that in our practical lives we have no alternative but to live from conjecture to conjecture. Like Abraham, the father of faith, as he has been called, we journey without knowing for certain where we will arrive.

Practical reason, like natural selection, is the tool of eliminating error. Scientific and religious conjectures eventually become refutable. The positivists were in error to regard religion to be mere private utterances and meaningless nonsense. Unification view describes religion as arising from the search for internal truth, for the purpose of moving from inner ignorance to inner knowledge. As humans have a body as well as mind, human search for truth includes science as well as religion. The object of scientific pursuit is external truth in order to move mankind from outer ignorance to outer knowledge. So, science and religion have in common that they share the pursuit of truth. We support the premise that it is part of man's nature to try understand his world and his rational attempt at it is an essential part of the theist's endeavor to examine the relationship between man, God and the world. Arising out of the scientists' exploration of the world is a sense of awe and respect and wonder that is akin to the religious sense. All this moves us to describe science and religion as one of the greatest spiritual adventures that man has yet known.

Theological statements, if satisfactorily formulated and related to a particular problem situation, say about the relationship between God, man and the world, can be rationally and critically discussed, with particular regard for their fruitfulness although truth cannot be established. Recognition that all our judgments are human and therefore fallible is not a denial of the objective nature of absolute truth. Truth remains the goal of human inquiry. An alteration in scientific outlook—for instance, the suppression of pure Newtonian Mechanics by relativity—is generally looked on as a victory for science, but an alteration in religious outlook—for instance, the abandonment of belief in the literal truth of the account of creation in Genesis—is usually looked on in some way as a defeat for religion. Yet either both are defeats or both victories. We can't give up the idea of ultimate sources of knowledge, and hold that all knowledge is human; that it is mixed with our errors, our prejudices, our dreams, and our hopes; that all we can do is grope for truth even though it is beyond our reach. I sincerely hope we will, through our enthusiastic debate, have a great success in this Unification Thought session in conjunction with IIFWP Assembly 2003. Thank you very much.

UNIFICATION OF SCIENCES AND UNIFICATION THOUGHT

KEYNOTE SPEECH AT THE 14TH INTERNATIONAL SYMPOSIUM ON UNIFICATION THOUGHT

DECEMBER 1, 2002, HOTEL METROPOLITAN, TOKYO, JAPAN

Most honorable and distinguished scholars, respected guests of this symposium on Unification Thought under the theme of The Unity of Sciences and Unification Thought, Ladies and Gentlemen:

It is a distinct honor for me to address you the topic Unification of Sciences and Unification Thought at the beginning of this very significant international symposium. In order to be here, you have taken time from your work and important schedules. I know this indicates your awareness that our world faces many serious challenges, and that it is our responsibilities to work together in the search for solutions to the critical problems we face today.

I.   The Ideal of Unity of Science

Before mentioning the unity of science, we should be agreed on the definition of the word 'science.' As in earlier times we identify science with knowledge, that is, knowledge of facts and theory. It covers the natural sciences and the social sciences, as well as a part of humanistic studies. In all the sciences, the reach is for truth or true knowledge. As to the unity of the sciences, all the sciences have a unified goal in the pursuit of truth. Probably scientists have from the beginning of systematized science had the thought that it would be necessary for them to find some kind of unity, and for that purpose they have also made some kind of code of science.

The unity of science is surely an ideal which is present, consciously or unconsciously, in the heart of every scientist. The unity of science appeared as an ideal to strive for when physics and chemistry developed with immense rapidity in the 17th century. Newton already united the descriptions of two phenomena: the motion of objects falling to ground, with the revolution of the moon around our Earth and the planets around the sun. The successes of the physical sciences raised the hope that ultimately the behavior of all objects could be described by the laws which can be obtained through the study of material objects.

After the Newtonian success, another profound unity established by modern physics was that of Einstein's relativity theory. A moving body becomes older at a slower pace than a resting body of the some kind. Space and time are thus intimately connected to the point of mutual convertibility at a fixed rate, which is measured by the velocity of light in vacuo. This new unity is one inseparable field called space-time.

Then, the emergence of field theories led to a unified treatment of electricity, magnetism, and light; quantum mechanics may be viewed as a formalism unifying the corpuscular and wave-packet aspects of physical reality. That part of quantum mechanics is referred to, which can be based on Schrödinger's equation, consistently with our knowledge of cosmic phenomena. Then, is there full unity in present day physics which deals, not only with phenomena which we can realize here on earth, but tries to account for events in the world at large, in the cosmos? Until now, the answer is 'no.'

The events between which Schrödinger's equation, that is quantum mechanics, describes correlations are fundamentally different from those of global physics, as embodied in the general theory of relativity. The events between which the theory of relativity establishes regularities are space-time coincidences. But, for microscopic physics, that is, quantum mechanics, space-time coincidences do not exist. Nevertheless, the physical scientist still goes in search of the unity or theoretical consistency between macroscopic global physics and microscopic quantum physics.

Beyond the physical sciences, the question of coherence within the sciences of psychology, anthropology, and also philosophy, in the Hellenistic era, is not easy to discuss. These fields are very different from the physical sciences. But I would like to present one model case based on Schrödinger. It appears that there is no role for life or consciousness within the field of the physical sciences. But quantum physicist Schrödinger, who is also one of the founders of molecule biology, searched for the problem of 'life' in respect of quantum mechanics. So, we could conclude that a greater unity of science is demanded not only by the present state of physics, but also by our general conceptions of science.

Moreover, I should like to stress the fundamental continuity or affinity between the intellectual man and the intelligible world, 'dead matter,' on one side and a 'divine soul' on the other. Matter is not stuff, and the spiritual is not ghostly. The former corresponds closer to the intensive, the latter more to the extensive dimensions of the world. We thus have to distinguish, but we must not separate. There will be some basic coherence. Science begins its research on visible and external things. However, science can also assist in understanding the frontiers of the invisible or internal things of the spiritual dimension. Thus, we must be able to have a central point located in the external visible world connected to the internal dimension. Together they will revolve around each other in eternal give-and-receive action.

Rev. Moon asserted at the ICUS conference, "Today, in all fields people are prone to narrow their research to small and limited areas so that they tend to lose the overall purpose or the centrality of their subject. In order to integrate all the specific fields of research, we are in need of a larger design or blueprint. In this way we may have a common ideal before us as we proceed to achieve this integration. The main purpose of this symposium on ICUS is for us produce that blueprint. Man is aware that he has life within him because his mind that originates from a source of the highest dimension is not limited to space and time. That source may be called true love or the Shimjung (心情) of the first cause of all beings. Man must be able to understand the centrality of absolute love in the cosmos in order to give his life meaning. We must not lose the very central point of the whole purpose."

II.   Unity of Science in the Vienna Circle

The atomists were the first reductionists. Democritus' atomism, certainly very popular in his day, was plausible, not because of the mechanistic postulation of atomic discontinuities, but because of the assumption that a very small stock of accessible concepts like shape, size, and speed could in principle explain all change and all diversity in the natural world. It assumed that the most complex properties of life and mind could, in principle, be understood in simple quantitative terms.

The success of Galilean mechanics was the second attempt at reductionist materialism. In his view, Secondary qualities (like color and viscosity) could be explained in terms of Primary qualities (like extension and mobility, which warranted the application of mathematized mechanics). Thus the language of science would constitute no problem since it would ultimately reduce to a set of quantitative empirical concepts. In this view, the unity of science must be attainable because all sciences ultimately reduce to the most general science, physics, i.e., physicalism.

Such concepts would constitute the universal principle of the unity of science in the Vienna Circle. By making science a single enterprise, epistemologically speaking, the unity of science could be achieved. In the 1930's, an Institute for the Unity of Science was set up under positivist auspices. Its slogan, 'Unity of Science (Einheitswissenschaft)' was created by that living dynamo, O.Neurath, and was taken up by other members, in particular R.Carnap, H. Reichenbach, C.G. Hempel, P.Frank, the physicist P.Bridgman, the biologist G.Wald, the statistician R.von Mises, the logician W.V.Quine, psychologist B.F.Skinner, young linguist, N. Chomsky and various sympathizers of logical empiricism in England, Germany, Austria, Poland, France, and the United States. Each of whom was one of the foremost thinkers in his field and among the most creative minds of the era, and all supporting that unity of science.

They began publication of an 'International Encyclopedia of Unified Science' which would lay the foundation for a new methodologically unified approach to cognitive assertion generally. Things quickly began to fall apart. In 1953, with the publication of Philosophical Investigations, a different Wittgenstein from the one who had authored the Tractatus—a model of logical empiricism—became known to the world and soon caught the fancy of philosophy and the other fields of science. Moreover, logical empiricism came under increasingly severe attacks from all sides, mainly, from Popper and his followers.

The original thesis of the unity of science was based on the belief that all scientific concepts are definable in the strictest sense of language of sense-data by terms belonging exclusively to the positivistic language or, alternatively, the physical language. The question whether intensional and other non-extensional logics are reducible to extensional ones—a positive answer to which was for Wittgenstein and Russell a matter of deep logical conviction—was dealt with by the later Carnap. He originally formulated the thesis of extensionability as one of the mainstays of the Unity of Science. I myself translated Meaning and Necessity, one of Carnap's later works, into Korean, and it mainly deals with reducibility, and the formulation of extensional logics.

Today these claims of reducibility look rather naive. Most logical empiricists came to realize that most terms occurring in scientific theories are neither definable by, nor reducible to, the so-called observational terms of the everyday thing-language. The Institute for the Unity of Science has recently dissolved. That its aims have been achieved might well be doubted. The original positivist manifesto has been abandoned. Though many would still set up a methodological unity of science as a goal, the barriers to the unification of the natural and the social sciences are serious ones. These are matters of the most intense controversy among philosophers of science at present.

The word science was taken in the wide sense of disciplined inquiry that includes all of the natural and social sciences and the humanities. The study of science itself requires close attention to the various inter-theoretical relationships; Chemistry is more than physics, biology is more than chemistry, psychology needs brain plus mind. Of course, the problems of reducing biology to physics, psychology to biology, or linguistics to psychology, though not considered by Carnap, etc., in their modern formulations, are intellectually more exciting than ever. If science is viewed as an ensemble of theories, the problem of the unity of science has traditionally been dominated by the idea of reduction, that is, upward causation, which means that theories about a higher ontological level can be reduced to a theory about a lower ontological level, which is the basis of the final cause.

However, contrary to reductionism, downward causation is also important. For example, knowledge about the cellular control of molecular activities may complement knowledge about the molecular control of cellular activities; that is, the higher cellular level controls the lower molecular level. As shown by N. Chomsky, the facts of language acquisition by children are better accounted for by assuming a special innate faculty of language based on mentalism than by relying on behavioristic learning theory.

Through the movement of Unity of Science mentioned briefly above, we know that scientists have tried to carry out the task and ideal of science for a long time. To sum up the history of that movement from the viewpoint of Unification Thought, the first try to realize the ideal of Unity of Science along with the developing science was the movement of the School of Encyclopedia based on the philosophy of enlightenment in the seventeenth Century. They compiled an encyclopedia on the basis of empiricism of the philosophy of enlightenment and mechanical materialism, which is the predecessor of The Encyclopedia Britannica enhancing its reputation even today.

The second attempt was the movement of Unity of Science centering on Vienna Circle mentioned above. Two movements for the Unity of Science with the development of science did not play the leading role in realizing the unity of science. If I point out the causes of the failure from the standpoint of Unification Thought, it is that the philosophical foundation of the two movements was put on empiricism and materialistic philosophy. Rev. Moon introduces God to us using a metaphor as 'Scientist of Scientists, and Artist of Artists', which emphasizes the importance that the foundation of science, holding theistic presuppositions, should be placed on the basis of God and Godism, because all natural things and the practical world are created by God. Therefore, any science or scientific research cannot get its justification unless it is based on God. It is why, spending an enormous expense, Rev. Moon have established ICUS with intellectual elites in the world and proposed the quest of science centering on Absolute Value and the movement of Unity of Science.

III. Holistic Worldview of Sciences

The view of unification thought is neither dualism, nor spiritualism, nor materialism. It is unitism or a theory of oneness. Unification thought says that all being, resembling the Original Image, is the united body of sung sang (mind) and hyungsang (matter). sung sang and hyungsang must share something in common in order to perform give-and-receive action between themselves. Descartes's dualism doubted that two essentially different elements could assume a mutual interaction, but we can definitely say that there is a hyungsang element in the sung sang, and a sung sang element in the hyung sang. Their relationship is that of subject and object, i.e., dominating and dominated, with one taking the controlling and active role, and the other, the obeying and passive role. There is no harmonious give-and-receive action without the central role of Shimjung or true love. True love brings a united body between the subject and the object.

Following unification thought, I will survey the holistic view of the sciences. Holistic theory describes the world as an undivided whole. First, the forces of molecules, the entelechy of organisms, the soul of animals, and the spirit of human beings possesses collective qualities somehow related to those we have encountered in quantum theory, which is a profoundly general science indeed. Quantum theory does not contain any antinomies. But there are startling paradoxes in view of the usual ontology with its strict regard for the separation of matter and mind, of the objective and the subjective.

In quantum mechanics, there are, everywhere, a great number and hierarchy of wholes that exceed the sum of their respective parts. This lesson has been dramatically exemplified by the quantum mechanical paradox of Einstein, Podolsky Rosen. They demonstrated in 1935 that this state of affairs is consistent with laws of quantum theory: A compositive system $\Sigma = \Sigma_1 \cdot \Sigma_2$ is in a maximally defined state:

$$\Psi = \sum_\alpha \Psi_1{}^\alpha \otimes \Psi_2{}^\alpha$$

wherein the physical system Σ is a whole unity which can be divided, by force, into its parts Σ_1 and Σ_2. But the proprietal state Ψ of Σ is an indivisible unity which cannot be separated intellectually into separate parts Ψ_1 and Ψ_2, in spite of the fact that the physical parts Σ_1 and Σ_2 are spatially and dynamically separated.

This remarkable individuality of atomic phenomena is not restricted to the smallest particles only; According to quantum theory, it applies to unities like molecules or organisms in an extremely interesting manner. The dynamical objectivity is combined with the informational subjectivity of this basic concept Ψ of a probability amplitude. Physics and epistemology, i.e., objectivity and subjectivity, become essentially identical in the context of quantum theory.

Classical physics saw the world as a regular, deterministic clockwork, ruled by Newtonian laws. This view changed dramatically, even as far as classical mechanics was concerned, with the discovery of chaotic systems. These systems are so sensitive to small perturbations from outside that they cannot be meaningfully separated from the rest of the universe and their behavior cannot be predicted in detail. Prigogine showed that behind the chaos there is a new type of order. It is a spontaneous order, the order exemplified, for instance, by the delicate balance of regularity.

Current mind-brain theory no longer dispenses with conscious mind as just an 'inner aspect' of brain activity, or as some passive epiphenomenal, metaphysical, or other impotent by-product, as has long been the custom; nor does it reject consciousness as merely an artifact of semantics or as being identical to the neural events associated with it.

Consciousness, in these revised terms, becomes an integral, dynamic property of the brain process itself and a central constituent of brain activity. Subjective experiences is viewed in R.W.Sperry's operational terms, as a causal determinant in brain function and acquires emergent control influence in regulating the course of physical-chemical events in brain activity. In a sense, mind moves matter in the brain just as an organism controls its component organs and cells, or a molecule governs the molecular course of its own electrons. The conscious mind as reinstated in the brain of objective science and scientific theory is squared with common sense on the mind-controlling-behavior issue.

IV. Sciences and Values

Facts and values are two categories not easily related. There is always a logical as well as a real gap between is and ought: between demonstration and vindication. General acceptance of the inadequacy of science in the realm of ethics and moral judgments is reflected in the old adage that 'science deals with facts, not with values.' and its corollary that 'value judgments lie outside the realm of science.' In other versions it is stated that science may tell us how but not why or that science may show us how to achieve defined goals, but not what those goals should be; science describes but cannot prescribe.

It seems that science as a discipline must by its very nature operate in the realm of objective fact and that science, as a method, can neither formulate value standards nor resolve issues in the domain of subjective value. This traditional separation of science and values and the related limitations this has implied for science as a discipline are no longer valid in the context of current mind-body theory. Human values can also be viewed objectively in scientific terms as universal determinants in all human decision-making. All decisions boil down to a choice among alternatives of what is most valued.

The importance of value issues is apparent also in another area, the so-called brain problem. The human brain comes equipped in advance with established value determinants and with inbuilt logical constraints that have their origins partly in biological heritage, partly in prior experience and may even arise through formal acceptance of ethical axioms. In practice, therefore, it is not a question of deriving values from the facts per se.

Incoming factual information interacts as a co-function with intrinsic cerebral value determinants in the building of one's sense of value. Along the above lines, progress could be greatly speeded on many fronts if we can clearly recognize that science deals with values as well as with facts.

Any given brain will respond differently to the same input, and will tend to process the same information in quite diverse ways depending on its particular value system. In short, what an individual or a society values determines very largely what it does. Values and other mental phenomena, though built of neural events, are no longer conceived to be reducible to, nor identifiable neither with, those events, nor to mere paralleled correlates.

This view of the unity of science and knowledge does not simply reduce all knowledge to a single academic discipline. Instead, it is a unity grounded in a fundamental purpose. By attempting to be value-neutral, science has often excluded the questions of humanity and moral values in the progress of its development. Another reason for the demise of value and morality is that past standards of value and morality no longer satisfy modern minds. New reasonable morals and ethics must arise out of a new standard of values. This standard can be derived only from a transcendent, unified system of thought that unifies science.

Rev. Moon has said, "I view the original character of science as embodying in unity the two sides of spirit and body, resembling man. This means that science should assume a unified character dealing also with the field of moral value. Therefore, to establish a true standard of value for the common benefit and welfare of mankind we cannot but set up as the standard a universal and absolute element that can be the central purpose of all humanity."

The essence of this absolute standard should be true love, which forms the basis of the system of ethics and values. It also forms the basis of all existing beings. I believe that this direction and standard of value can come only from God-centered thought. I maintain that the absolute values we seek are grounded in the absolute true love of God. Solutions to the world's problems can come about only through this holistic approach to human existence.

V.   Logical Foundation for the Unity of Science

Briefly, I'd like to introduce unification logic in order to search for the foundation of the unity of science. Logic and mathematics are clearly very different from the other sciences: they appear to describe, or be based on abstract arguments. It is generally admitted, that they act intuitively based on their application in the other sciences—which does serve marvelously. Logic is concerned with laws of human thinking and scientific reasoning.

Thinking proceeds in the direction of realizing the purpose of human beings, and the purpose is fundamentally rooted in the Heart, or Shimjung in
the view of unification thought. Because thinking has a direction towards
the object, cognition and thinking necessarily involve the subject's action
towards the object in order to realize its purpose.

The logical structure of human thinking resembles the logical structure
of the Original Image in principle. The direction to realize the purpose
rooted in the Heart shapes a plan through thinking resulting from the mutual relationship and unity of the two elements, that is, intention(the directional structure of the rational faculty) and its corresponding concept. The
logical structure of this four position foundation is not formed by experience but through an innate, a priori process, although experience is not
wholly excluded in its making. In fact, without experience of daily life, we
cannot explain the interaction of the logical structure and the practical
structure.

Aristotle's attempt is the first to construct a system of all sciences based
on formal logic. Science in his view must rest on first principles, themselves necessary truths, certified as true by the skilled insight of the
philosopher-scientist. From these as premises, deductive logic can generate
demonstrative science, in which the more specific conclusions appear as
necessary truths. Lacking such principles, there is no way in which an eternal and necessary science could be constructed.

Aristotelian logic provided a tight deductive and conceptual unity of
science, since all conclusions have to be implicit within the premises with
which the deductive inference begins. Particularly, the formal logic of Aristotle has been regarded as self-evident truth. But, is it really so? If logic
aims at setting the foundation for a proper scholarship of the sciences, it
must be admitted that the foundation of the 'formality' in formal logic is
weak. This is because we cannot assert the universality of the sciences
from the abstract 'formality,' which originated in empirical positivity.

Unification Logic recognizes the a priori sphere of subjective conditions
as the foundation of the possibility of such positive experiences and as that
which controls all empirical conditions. Accordingly, from the viewpoint of
Unification Logic, the foundation for proper sciences can begin neither
with empirical 'positivity' alone nor with the subjective a priori alone. It is
Unification Logic, which is being promoted as an alternative to formal
logic, which has been the basis of most academic endeavors.

As seen above, Aristotle's deductive logic could not become the universal principle to establish the foundation of metaphysics and science. From
the viewpoint of Unification Thought, the system of formal logic of Aristotle is literally formal system, which is a vacant logic system to tell nothing about the empirical essentialities and empirical science, which can become the basis of science. To establish an eternal and necessary foundation, as Aristotle mapped out, the deductive motive coming from the mind,

sphere of sung sang, and the inductive motive of the level of experience, that is, sphere of hyungsang should be under consideration. In Unification Thought, when these two motives, that is, deductive motive and inductive motive, are under consideration in the sense of subjective and objective, the basis of true science can be secured.

Kant saw no way of explaining the universality and necessity of the law of Newtonian mechanics, Other than that he supposed that the source of this universality and necessity lay in the human mind itself. As seen from Hume's inductive argument, no particular sense could of itself achieve the certitude of universality and necessity. The mind constructs in advance, for example, the categories of a 'pure physics' of synthetic a prior truth. Kant was forced to look carefully at the mind's power of projecting its own structures into the objects of scientific inquiry.

From the viewpoint of Unification Thought, the typical example of the universal principle which can be a basis of science is the principle of inter-action of give and receive. Science is a process of seeking truth to try to explain an individual experienced facts through a general and universal principle. Accordingly, the work to establish science cannot be achieved in the sphere of experience nor in the sphere of mind either. Because the di-mension of mind and the dimension of experience is matching in the recip-rocal relationship of subjectivity and objectivity.

So, neither of them can regulate the other. The unique principle to work between both parties is the principle of give-and-receive. If you seek for the principle of regulation in mind only, the empirical world is nothing but a chaotic various collection of sense data. If that is so, isn't the mechanic law of Newton applied in this world realistically? Are all those laws sub-jective mind laws?

Unification Epistemology maintains that man's subjectivity is involved in cognition. Of course, the sensible qualities of the object, also, are neces-sary for cognition to take place. Accordingly, as the general principle that science seeks for appears in the mind as the law of presentation and practi-cal law in the world of experience, which are regarded to have a relation-ship of subjectivity and objectivity. From the viewpoint of Unification Thought, our subjectivity is able to have cognizance of the physical law of the practical world for the first time. In this respect, Unification Thought stands above Kant.

May your intense discussions during this symposium be successful and most fruitful. I pray for God's protection and blessing to be with you and your families and your nations.

Thank you very much.

UNITY OF ACADEMIC DISCIPLINES AND UNIFICATION THOUGHT

KEYNOTE SPEECH AT UNIFICATION THOUGHT SESSION, WCSF 2003

JULY 11, 2003, SUN MOON UNIVERSITY, ASAN, KOREA

Most honorable and distinguished scholars, respected guests of this symposium on Unification Thought, Ladies and Gentlemen: It is a distinct honor for me to address you the topic Unity of Academic Disciplines and Unification Thought at the beginning of this very significant international symposium. In order to be here, you have taken time from your work and important schedules. I know this indicates your awareness that our world is faced with many serious challenges, and that it is our responsibility to work together in a search for solution to the critical problems we face today.

I.   Two Types of Unity of Science

A) The Type of Causation—Reductionism

Through the movement of unity of science, we know that scientists have tried to carry out the task and ideal of science for a long time. To sum up the history of that movement from the Unification Thought view, the first attempt tried to realize the ideal of Unity of Science along with the developing science, which was the movement of the School of Encyclopedia based on the philosophy of enlightenment in the seventeenth century. They compiled an encyclopedia on the basis of empiricism of the philosophy of enlightenment and mechanical materialism, which is the predecessor of The Encyclopedia Britannica enhancing its reputation even today. The second attempt was the movement of Unity of Science centering on the Vienna Circle. The academic trend in the School of Encyclopedia or the Vienna Circle was originated by the tradition of empirical epistemology and materialistic philosophy, and the two perspectives were strong advocates of empirical reductionism. I will call these traditions "the Type of Upward Causation."

This type of upward causation had a tendency to base epistemological foundation of all sciences on empirical sense data and to base ontological foundation of all sciences on materialism. The atomists were the first reductionists. Democritus' atomism, certainly very popular in his day, was plausible, not because of the mechanistic postulation of atomic discontinuities, but because of the assumption that a very small stock of accessible concepts like shape, size, and speed could in principle explain all change and all diversity in the natural world. It assumed that the most complex properties of life and mind could, in principle, be understood in simple quantitative terms. The success of Galilean mechanics was the second attempt at reductionist materialism. In his view, secondary qualities (like color and viscosity) could be explained in terms of primary qualities (like extension and mobility, which warranted the application of mathematized mechanics). Thus the language of science would constitute no problem since it would ultimately reduce to a set of quantitative empirical concepts. In this view, the unity of science must be attainable because all sciences ultimately reduce to the most general science, physics (physicalism). Such concepts would constitute the universal principle of the unity of science in the Vienna Circle.

By making science a single enterprise, epistemologically speaking, the unity of science could be achieved. The original thesis of the unity of science was based on the belief that all scientific concepts are definable in the strictest sense of language of sense data by terms belonging exclusively to the positivistic language or, alternatively, the physical language. The question whether intentional and other non-extensional logics are reducible to extensional ones to which were for early Wittgenstein and Russell a matter of deep logical conviction, which was dealt concretely with by the Carnap. He originally formulated the thesis of extension-ability as one of the mainstays of the unity of science in a very calm and definite way. Today these claims of reducibility look rather naive. Most logical empiricists came to realize that most terms occurring in scientific theories are neither definable by, nor reducible to, the so-called observational terms of the everyday thing-language. The Institute for the unity of science has recently dissolved. That its aims have been achieved might well be doubted. The original positivist manifesto has been abandoned. Though many would still set up a methodological unity of science as a goal, the barriers to the unification of the natural and the social sciences are serious ones. These are matters of the most intense controversy among philosophers of science at present.

The Type of Downward Causation—Unificationism

The two movements for the unity of science through the development of science, above mentioned, did not play a leading role in realizing the unity of science. If I point out the causes of the failure of the two movements from the standpoint of Unification Thought, we could say that they were founded on a mechanistic world-view, or the premise of the positive philosophy, and that they advocated that premise dogmatically and intolerantly. The unification of science is a work that can never be attained from a dogmatic or intolerant initiative. Accordingly, the unification theory of science can be actualized when it is based on a more comprehensive and universal premise and philosophical foundation. It is for this reason that we need this kind of philosophy. If science is viewed as an ensemble of theories, the problem of the unity of science has traditionally been dominated by the idea of reduction, that is, upward causation, which means that theories about a higher ontological level can be reduced to a theory about a lower ontological level, which is the basis of the final cause.

In short, higher and more comprehensive scientific theory in the sphere of ontology should be reduced to lower and more basic scientific theory in the sphere of ontology. It is due to the belief that the language and scientific theory of the sensory data of physics, the lowest ontological dimension, is the fundamental cause of all studies. However, contrary to reductionism, downward causation is also important. As reductionism, upward causation is persuasive when an atomist explains a mechanistic world-view such as materialism and behaviorism, downward causation is also more persuasive in explaining an organic world-view of teleology such as idealism and spiritualism. For example, knowledge about the cellular control of molecular activities may complement knowledge about the molecular control of cellular activities; that is, the higher cellular level controls the lower molecular level. As shown by N. Chomsky, the facts of language acquisition by children are better accounted for by assuming a special innate faculty of language based on mentalism than by relying on behaviorist learning theory. Theory of N. Chomsky's linguistics presents a model of the typical downward causation through taking out language ability of man from the innate faculty his spirit obtained. From this viewpoint, Unification Thought is also considered a downward causation centering on God and spiritual value. In Unification Thought, the mind and body of a man is in unity, naming the former and the latter as subject and object, cause and effect, and vertical and horizontal relationship respectively. In this case, mind-body is a causal cause-effect, so the causal cause is established by the subjective mind.

Therefore, the subject-object relationship becomes a model of downward human relationship having the subject as a cause. For example, in a human body, the subconsciousness of life (mind) is the subject, and the organization and the cell of a human body becomes the object. Judging from the law of cause and effect, the subjective life (mind) becomes the cause to control and dominate the objective body. This subject-object relation model is not limited to the relationship of mind and body but becomes the most fundamental principle to control the existing world of man and nature. To say it once more, all beings existing in the world are a methodical organic body connected by causational chains of subject and object. The human world as well as the natural world is composed of a grand methodical unified body. This principle to maintain the methodical system of subject-object in nature is natural law, and the one that maintains the methodical system of subject and object in human relationships is the law of value or ethical law. Therefore, it is the task of all studies in science, philosophy, and arts and so on to research the laws controlling these two worlds. In this case, the subject holds priority over the object in the sphere of ontology and axiology. It is because the subject holds the subjective, affirmative, centripetal, and vertical value, that the object holds superficial, passive, peripheral, and horizontal value, to maintain the reciprocal relation. As there is no body without mind, there is no science without value. This is because the relationship between value and fact is the necessary relationship of the subject and object. It can be seen that the ideal of unity of science, which failed to actualize in the way of Reductionism (Upward Causation), can be actualized in the way of Downward Causation. It is, therefore, said that Unification Thought is a theoretical system that restores a real position of academic disciplines and truly unifies all academic disciplines by properly establishing the relationship of subject-and-object in the law of cause-and-effect.

II.   Unification Logical View in the Unity of Science

Aristotle's attempt is the first to construct a system of all sciences based on formal logic. Science in his view must rest on first principles, themselves necessary truths, certified as true by the skilled insight of the philosopher-scientist. From these as premises, deductive logic can generate demonstrative science, in which the more specific conclusions appear as necessary truth. Lacking such principles, there is no way in which an eternal and necessary science could be constructed. Aristotelian logic provided a tight deductive and conceptual unity of science, since all conclusions have to be implicit within the premises with which the deductive inference begins. Particularly, the formal logic of Aristotle has been regarded as self-evident truth. But, is it really so?

If logic aims at setting the foundation for a proper scholarship of the sciences, it must be admitted that the foundation of the 'formality' in formal logic is weak. This is because we cannot assert the universality of the sciences from the abstract 'formality' which originated in empirical positivity. Unification Logic recognizes the a priori (priority) sphere of subjective conditions as the foundation of the possibility of such positive experiences and of those which control all empirical conditions. Accordingly, from the viewpoint of Unification Logic, the foundation for proper sciences can begin neither with empirical 'positivity' alone nor with the subjective a priori alone. It is Unification Logic which is being promoted as an alternative to formal logic, which has been the basis of most academic endeavors. As seen above, Aristotle's deductive logic could not become the universal principle to establish the foundation of metaphysics and science. From the viewpoint of Unification Thought, the system of formal logic of Aristotle is literally formal system, which is a vacant logic system to tell nothing about the empirical essentialities and empirical science, which can become the basis of science.

To establish an eternal and necessary foundation in the view of Unification Thought, as Aristotle mapped out, the deductive motive coming from the mind, sphere of sung sang, and the inductive motive of the level of experience, that is, sphere of hyungsang should be under consideration. In Unification Thought, when these two deductive and inductive motives are under consideration in the sense of subjective and objective, the basis of true science (metaphysics and physics) can be secured. Kant saw no way of explaining the universality and necessity of the law of Newtonian mechanics, other than that he supposed that the source of this universality and necessity lay in the human mind itself. As seen from Hume's inductive argument, no particular sense could of itself achieve the certitude of universality and necessity. The mind constructs in advance, for example, the categories of a 'pure physics' of synthetic a priori truth. Kant was forced to look carefully at the mind's power of projecting its own structures into the objects of scientific inquiry. He believed that the mind is already equipped structurally with the categories of 'pure physics' meaning the synthetic a prior truth. Through that, he thought the universalities and necessities of empirical science could be regulated from the innate a priori structure of mind.

From the viewpoint of Unification Thought, the typical examples of the universal principle that can be a basis of science is the principle of interaction of give and receive. Science is a process of seeking truth to try to explain individual experienced facts through a general and universal principle. Accordingly, the work to establish science cannot be achieved in the sphere of experience or in the sphere of mind either. It is because the dimension of mind and the dimension of experience are matched in the reciprocal relationship of subjectivity and objectivity. So, neither of them can regulate the other. The unique principle to work between both parties is the principle of give-and-receive. If you search for the principle of regulation in mind only, the empirical world is nothing but a chaotic various collection of sense data. If that were so, wouldn't the mechanic law of Newton apply in this real world? Are all those laws subjective mind laws? Do we need a fundamental principle to regulate the experimental physical world? Unification Epistemology maintains that man's subjectivity is involved in cognition. Of course, the sensible qualities of the object, also, are necessary for cognition to take place. Accordingly, the general principle that science seeks for appears in the mind as the law of representation and practical law in the world of experience, which are regarded to have a relationship of subjectivity and objectivity. Those two subjective and objective categories include the relationship of subjectivity and objectivity, and the principle to regulate the spiritual world of mind and physical world of experience respectfully. From this viewpoint our cognition can obtain the ability to recognize the physical laws of the empirical world, and accordingly, embrace all the subjectivity and objectivity simultaneously. In this respect, Unification Thought stands above Kant. Furthermore, it is possible to establish the theory for the unity of sciences on the structure of the logical foundation of Unification Thought.

III.   Common Ground of Religion and Science

Central notion on the positivist theory of science is the concept of verification. As you know, positivists' presumptions have had a deep and lasting influence on the general intellectual climate of our time. We have seen that the verification principle as an extension of this thesis became an arbiter between sense and nonsense and therefore an essential criterion of rationality. Theological assertions, because unverifiable, are thus rendered nonsensical and therefore irrational. Consequently, theologians who opposed this verification in science, and its extension as a general principle of rationality, felt uneasy about the problem of verification in religion. But the theory of science is not the process of an inductive generalization to collect data and evidences through inductive work, but is a scientific thinking process to present hypotheses through a bold conjecture, intuition and imagination. Rather, science is the discipline which intensifies:

1. Imagination,
2. Critical discussion, and
3. Experimental testing in the quest for truth.

The point is that, without daring and bold conjectures and hypotheses that grow out of the attempt to solve cognitive problems, we could not improve our knowledge. Religion and science do not differ with regard to vision and imagination, for either will perish without vision. Belief is also a major part of the scientific process. If science were nothing but skepticism, nothing would ever come of it, that is, no scientific advances would be made. Belief is the tendency to act on the basis of expectations. When scientists test their hypotheses and conjectures on those expectations, they are then living up to the scientific ideal. So far, religious commitment and scientific commitment are seen to be psychologically very similar. Both require belief. But this leads to the question of the role of doubt in the form of severe critical discussion and crucial experimentation. It is sometimes said that science differs from religion in that science requires critical debate, whereas religion tend to resist it. But this is oversimplification. It would appear that in our practical lives we have no alternative but to live from conjecture to conjecture. Like Abraham, the father of faith, as he has been called, we journey without knowing for certain where we will arrive.

Practical reason, like natural selection, is the tool of eliminating error. Scientific and religious conjectures eventually become refutable. The positivists were in error to regard religion to be mere private utterances and meaningless nonsense. Unification view describes religion as arising from the search for internal truth, for the purpose of moving from inner ignorance to inner knowledge. As humans have a body as well as mind, human search for truth includes science as well as religion.

The object of scientific pursuit is external truth in order to move mankind from outer ignorance to outer knowledge. So, science and religion have in common that they share the pursuit of truth. We support the premise that it is part of man's nature to try understand his world and his rational attempt at it is an essential part of the theist's endeavor to examine the relationship between man, God and the world.

Arising out of the scientists' exploration of the world is a sense of awe and respect and wonder that is akin to the religious sense. All this moves us to describe science and religion as one of the greatest spiritual adventures that man has yet known.

Theological statements, if satisfactorily formulated and related to a particular problem situation, say about the relationship between God, man and the world, can be rationally and critically discussed, with particular regard for their fruitfulness although truth cannot be established. Recognition that all our judgments are human and therefore fallible is not a denial of the objective nature of absolute truth. Truth remains the goal of human inquiry.

An alteration in scientific outlook—for instance, the suppression of pure Newtonian Mechanics by relativity—is generally looked on as a victory for science, but an alteration in religious outlook—for instance, the abandonment of belief in the literal truth of the account of creation in Genesis—is usually looked on in some way as a defeat for religion. Yet either both are defeats or both victories. We can't give up the idea of ultimate sources of knowledge, and hold that all knowledge is human; that it is mixed with our errors, our prejudices, our dreams, and our hopes; that all we can do is grope for truth even though it is beyond our reach.

I sincerely hope we will, through our enthusiastic debate, have a great success in this Unification Thought session in conjunction with IIFWP Assembly 2003. Thank you very much.

Introduction to the Unification Thought Movement

Keynote Speech at the 15th International Symposium on Unification Thought

November 28, 2003, Aerostar Hotel, Moscow, Russian

Federation

I would like to start by giving my most heart-felt welcome to Dr. Sook, Dr. Petrovsky, honorable guests, participants, and to all ladies and gentlemen.

I.   Characteristics of Unification Thought

In the summer of 2001 Rev. Sun Myung Moon gathered world leaders and continental directors at Namwon Training Center for 2 weeks, in Jeju Island, Korea for the 1st International Workshop on Unification Thought and theory of Victory-Over-Communism. Usually when Rev. Moon gathers world leaders there is some kind of providential meaning. At that time, however, many of the leaders were puzzled as to why Rev. Moon would want to have a workshop on Unification Thought and VOC theory at this time. This workshop became a special occasion for us to study Unification Thought with Rev. Moon.

Unification Thought is intended to systematize the thought of Rev. Sun Myung Moon and present it in an appropriate order. Unification Thought aims to realize a world in which all of humankind can serve God as one great family and bring about a peaceful world. In this respect, it is meant to serve to bring about world of peace, liberation, and true love through a process of reconciliation and unification. This intent may provide insight into why Unification Thought is also referred to as Godism or Headwing Thought. Godism refers to the fact that this thought has God's truth and his love at its core, and Headwing Thought refers to the fact that it is neither right-wing nor left-wing in its approach, but instead embraces the two by considering them from a higher perspective.

Unification Thought, however, does not only intend to reconcile and unify democracy and communism, which have been locked in ideological confrontation, but is presented as a system that can offer solutions to confusion affecting the human and social sciences, the natural sciences, and the arts. The unity and harmony it brings to the various theories competing in these fields must be based on the fundamental principles of the universe.

This will mark the 6th year since the passing of Dr. Sung-Han Lee, who systematized Unification Thought. At the 9th symposium held at Sunmoon University, Dr. Lee passed away. Since then, UTI has held up to the 15th symposium, now in Moscow. The 11th was held in the Philippines. Over 50 university president-class scholars and 300 top-class scholars from various fields participated in the symposium which was co-hosted with the PWPA. Mrs. Arroyo, the current president and then vice-president of the Philippines, offered the congratulatory address during this symposium. The minister of education of the Philippines and then president of PWPA Philippines did a good job in organizing this symposium, making it a great success.

After that, the 12th symposium was held in Taiwan, and the 13th in the Czech Republic, which is in Eastern Europe. The 14th symposium was held in Tokyo, Japan in 2002. Besides this International Symposium on Unification Thought, the Unification Thought Institute has hosted the International Conference on the Unity of Science (ICUS). There was a breakout session that dealt with Unification Thought during ICUS. During the 14th ICUS in America in 1985, Rev. Moon gave instructions to create a ground-breaking session in ICUS where world leaders could discuss Unification Thought. Since then, many world scholars have been studying Unification Thought including this year, during the World Culture Sports Festival 2003. The Unification Thought has now reached the stage where prominent scholars are taking interest in its theories and is being verified by the world scholars.

Rev. Moon has stated that leaders of the Unification movement must be armed with at least three basic theories. They are the Divine Principle, Unification Thought, and VOC theory. The Divine Principle takes on the structure of Christianity's systematic theology and the Unification Thought and VOC theories take on a more philosophical and ideological contour but they all come from Rev. Moon's words. The Divine Principle is what we received when we organized Rev. Moon's words in the form of systematic theology. When we rearranged his words into a more ideological context we got the Unification Thought VOC theory. Therefore we can see that all these theories have come from the words of the Rev. Sun Myung Moon.

Religion all originated from one God. But in the historical course of their formation, each religion lost its original purpose and multiplied hatred and anger towards other religions due to the limitations of the age, dogmatism, and stubborn insistence on one's doctrines. The terrorist attacks on the World Trade Center, the war in Afghanistan and Iraq, and other forms of civilization clashes, all stem from religious conflicts. Therefore it is important that man break away from dogmatic religion created by man, break away from the prison of religious ideology, and return to the spirit of its origin: scriptures.

Unification Thought is also called Godism or Headwing Thought. But looking at it from its founder's point of view it can be called the Thought of Rev. Moon. Just as we call ideologies by its founder's name like Maoism and Marxism, this Thought can be rightly called Moonism. Everything written in the Unification Thought by Dr. Lee has been reported to Rev. Moon, which he revised, and improved. In particular, the entire chapter of epistemology was covered by Rev. Moon. The 'Theory of the Original Image' in the Unification Thought pertains to the views of God in Christianity and Rev. Moon himself also formulated these conceptions. Rev. Moon is the alpha and omega of the Unification Thought and he was the one who conceptualized his own words into ideological format. Thus the Unification Thought can be called Moonism.

What kind of person is Rev. Moon? I dare say, he is the perfect incarnation of truth. Rev. Moon, who is the perfect being with Logos, has now brought the truth of God that was hidden to light. Thus the Unification Thought is not empirical knowledge that was obtained from man's reason or some kind of deductive or inductive method. Therefore the Unification Thought is not a thought of this earth but is a thought that came from Heaven. In that light, this thought must not be conveyed in a logical, consistent, and persuasive manner but must be proclaimed with absolute authority as The Truth. Please bear in mind that Unification Thought is Heaven's Thought and the fundamental principle, which will solve all ideological problems. As Rev. Moon talked about the Unification Thought in one of his speeches, let us read from his speech at the International Rally held in Seoul on November 16, 1985.

"Unification Thought is a powerful key capable of solving any problem, no matter how difficult it may be. When this thought is applied to society, various social problems can be settled. When this thought is applied to the world, world problems can be realistically solved. And particularly, when it is applied to criticizing Communist theories and theories of evolution, all the contradictions of Communism and Darwinism are brought to light, and a counterproposal can be established. This Thought presents a new view of life, a new view of world, a new view of universe and a new view of God's work history. It is also a principle of integration that can bring different religious doctrines and philosophies into unity, while preserving their diverse characteristics. I call this Thought "Unification Thought, or Godism."

This has been Rev. Moon's basic stance in explaining the Unification Thought. Why is it that the Unification Thought can disclose the limitations of other philosophies, ethics, and religious doctrines and present an alternative solution that can solve the problems that could not be solved by those ideologies that emerged in order to solve those very problems? After learning the Unification Thought I have experienced many times academic excitement coming from seeing these unsolved problems being solved through the Unification Thought. If you immerse yourself into the Unification Thought, you will be swept away by this ideological passion.

How indeed can I give a introduction on Unification Thought in this short given time? How can I comment on the philosophies of Kant, Hegel, and Plato and talk about the greatness of the Unification Thought during this time? Still I would like to mention the conclusion I came after studying Unification Thought : The Unification Thought is a great ideology. The basic concepts of the Unification Thought are simple. Sungsang, Hyungsang, Yang and Yin, Give-and-Receive Action, 4-position base etc. But these are the keys that will solve those fundamental problems that any thoughts so far could not solve. As a person who studies these thoughts, how can you feel so tremendous pleasure and joy?

II.   Unification Thought as the Key to Solving Actual Problems

We can see that the Unification Thought has the function of healing that will solve the actual problems which modern-day society faces. Unification Thought is not a theory for a theory but the key to the solution to actual problems. Today, living in this age, we sometimes experience the savage side of man. It is just like especially when we see the terror attacks, religious and racial strife among men. The family, youth problem in our society has reached a serious point. As you all know, we have access to unimaginable erotic pictures just a few clicks away on the Internet. Our young people are exposed to this kind of obscene Internet environment and cyber space.

Today, many people try to solve the conflicts and violence in this world with force. The Iraq war is a good example of this rationale. But in the end this method must rely on the theory of force whether it is political, economical, and military might. Violence leads to even greater violence. Mutuality is the basic characteristic of violence. Thus violence only breeds to greater violence. In other words we cannot find the ultimate solution to our problems through violence.

The conflict between Islam and Christianity is like that of a fight between brothers. They both believe in Abraham as their ancestor of faith. Thus we can regard their struggle as the struggle between Abraham's two sons, Isaac and Ishmael. It is the only parents who can solve this struggle between brothers. Therefore the way for the peaceful resolution to these problems can be accomplished only through heart, love, and reconciliation. Isn't that so? It is the same with the ideological problems. we cannot solve the conflicts and struggles in this world without a philosophy based on heart and love. That is why the ultimate solution to these problems we face today can only be solved through philosophy of true love, namely the Head-wing Thought or Godism. This is the only alternative.

The Cold War has not ended especially in the Korean peninsula, and the nation is still at a face-off due to ideological conflict. That the Soviet Union, a communist state has collapsed does not necessarily mean that the substance of communism has gone away. Rev. Moon is the only one who can save communism and lead it towards the right path. Ultimately we must solve all the religious, racial problems, the problem of North and South division, ideological differences, the family problem, and religious conflicts with the Unification Thought which will become our strong weapon in solving the world problems. It is not until a new dawn approaches that the darkness of night dilutes. Thus when the light of God's thought and truth is revealed, the world of darkness will soon be dispersed.

I sincerely hope that this symposium does not end in the exchange of the opinions of various scholars, but by applying Unification Thought to your own area of specialization, a turning point for forming and reconstituting new science will be achieved. When this movement influences the scholarly world peace in the era of global village will come even more quickly. It is my hope that through this symposium, Unification Thought will be developed into a movement of sciences. In this sense, this symposium builds up the flow of a new science with Unification Thought as the centripetal point. The papers to be presented this time are new in terms of their composition and creativity, and each bears the hard work of the presenters. I would like to request all of you to show keen interest in them and hold lively discussions on them. Moreover I truly hope at this moment, the meeting of the oriental and occidental scholars centering on Unification Thought becomes the cornerstone of a peace movement. Thank you very much.

Introduction to the Unification Thought Movement

Keynote Speech at the 16th International Symposium on Unification Thought

November 26-29, 2004, Crystal Palace Hotel, Sofia, Bulgaria

I. Characteristics of Unification Thought

I would like to start by giving my most heart-felt welcome to Rev. Song Young Cheol, Dr. Kaltchev, President Kyung-Jone Lee, honorable guests, participants, and ladies and gentlemen.

Unification Thought is intended to systematize the thought of Rev. Sun Myung Moon and present it in an appropriate order. Unification Thought aims to realize a world in which all of humankind can serve God as one great family and bring about a peaceful world. In this respect, it is meant to serve to bring about world of peace, liberation, and true love through a process of reconciliation and unification. This intent may provide insight into why Unification Thought is also referred to as Godism or Headwing Thought. Godism refers to the fact that this thought has God's truth and his love at its core, and Headwing Thought refers to the fact that it is neither right-wing nor left-wing in its approach, but instead embraces the two by considering them from a higher perspective.

Unification Thought, however, does not only intend to reconcile and unify democracy and communism, which have been locked in ideological confrontation, but is presented as a system that can offer solutions to confusion affecting the human and social sciences, the natural sciences, and the arts. The unity and harmony it brings to the various theories competing in these fields must be based on the fundamental principles of the universe.

This will mark the 7th year since the passing of Dr. Sang-Hun Lee, who systemized Unification Thought. At the 9th symposium held at Sunmoon University, Dr. Lee passed away. Since then, UTI has held up to the 16th symposium, now in Sofia in Bulgaria. The 11th was held in the Philippines. Over 50 university president-class scholars and 300 top-class scholars from various fields participated in the symposium which was co-hosted with the PWPA. Mrs. Arroyo, the current president and then vice-president of the Philippines, offered the congratulatory address during this symposium. The minister of education of the Philippines and then president of PWPA Philippines did a good job in organizing this symposium, making it a great success.

After the 12th symposium was held in Taiwan, and the 13th in the Czech Republic, which is in Eastern Europe. The 14th symposium was held in Tokyo, Japan in 2002. Besides this international Symposium on Unification Thought, the Unification Thought Institute has hosted the International Conference on the Unity of Science (ICUS). There was a breakout session that dealt with Unification Thought during ICUS. During the 14th ICUS in America in 1985, Rev. Moon gave instructions to create a ground-breaking session in ICUS where world leaders could discuss Unification Thought. Since then, many world scholars have been studying Unification Thought including this year, during the World Culture Sports Festival 2003. The Unification Thought has now reached the stage where prominent scholars are taking interest in its theories and is being verified by the world scholars.

Rev. Moon has stated that leaders of the Unification movement must be armed with at least three basic theories. They are the Divine Principle, Unification Thought, and VOC theory. The Divine Principle takes on the structure of Christianity's systematic theology and the Unification Thought and VOC theories take on a more philosophical and ideological contour but they all come from Rev. Moon's words. The Divine Principle is what we received when we organized Rev. Moon's words in the form of systematic theology. When we rearranged his words into a more ideological context we got the Unification Thought VOC theory. Therefore we can see that all these theories have come from the words of the Rev. Sun Myung Moon.

Religion originated from one God. But in the historical course of their formation, each religion lost its original purpose and multiplied hatred and anger towards other religions due to the limitations of the age, dogmatism, and stubborn insistence on one's doctrines. The terrorist attacks on the World Trade Center, the war in Afghanistan and Iraq, and other forms of civilization clashes, all stem from religious conflicts. Therefore it is important that man break away from dogmatic religion created by man, break away from the prison of religious ideology, and return to the Spirit of its origin: scriptures.

Unification Thought is also called Godism or Headwing Thought. But looking at it from its founder's point of view it can be called the Thought of Rev. Moon. Just as we call ideologies by its founder's name like Maoism and Marxism, this Thought can be rightly called Moonism. Everything written in the Unification Thought by Dr. Lee has been reported to Rev. Moon, which he revised, and improved. In particular, the entire chapter of epistemology was covered by Rev. Moon. The `Theory of the Original Image' in the Unification Thought pertains to the views of God in Christianity and Rev. Moon himself also formulated these conceptions. Rev. Moon is the alpha and omega of the Unification Thought and he was the one who conceptualized his own words into ideological format. Thus the Unification Thought can be called Moonism.

What kind of person is Rev. Moon? I dare say, he is the perfect incarnation of truth. Rev. Moon, who is the perfect being with Logos, has now brought the truth of God that was hidden to light. Thus the Unification Thought is not empirical knowledge that was obtained from man's reason or some kind of deductive or inductive method. Therefore the Unification Thought is not a thought of this earth but is a thought that came from Heaven. In that light, this thought must not be conveyed in a logical, consistent, and persuasive manner but must be proclaimed with absolute authority as The Truth. Please bear in mind that Unification Thought is Heaven's Thought and the fundamental principle, which will solve all ideological problems. As Rev. Moon talked about the Unification Thought in one of his speeches, let us read from his speech at the International Rally held in Seoul on November 16, 1985.

Unification Thought is a powerful key capable of solving any problem, no matter how difficult it may be. When this thought is applied to society, various social problems can be settled. When this thought is applied to the world, world problems can be realistically solved. And particularly, when it is applied to criticizing Communist theories and theories of evolution, all the contradictions of Communism and Darwinism are brought to light, and a counterproposal can be established. This Thought presents a new view of life, a new view of world, a new view of universe and a new view of God's work history. It is also a principle of integration that can bring different religious doctrines and philosophies into unity, while preserving their diverse characteristics. I call this thought "Unification Thought or Godism."

This has been Rev. Moon's basic stance in explaining the Unification Thought. Why is it that the Unification Thought can disclose the limitations of other philosophies, ethics, and religious doctrines and present an alternative solution that can solve the problems that could not be solved by those ideologies that emerged in order to solve those very problems? After learning Unification Thought I have experienced many times academic excitement coming from seeing these unsolved problems being solved through the Unification Thought. If you immerse yourself into the Unification Thought, you will be swept away by this ideological passion.

How indeed can I give an introduction on Unification Thought in this short given time? How can I comment on the philosophies of Kant, Hegel, and Plato and talk about the greatness of the Unification Thought during this time? Still I would like to mention the conclusion I came after studying Unification Thought: The Unification Thought is a great ideology. The basic concepts of the Unification Thought are simple: suugsang, hyungsang, yang and yin, Give-and-Receive Action, 4-position base, etc. But these are the keys that will solve those fundamental problems that any thoughts so far could not solve. As a person who studies these thoughts, how can you

feel so tremendous pleasure and joy?

II.   Unification Thought as the Key to Solving Actual Problems

We can see that Unification Thought has the function of healing that will solve the actual problems which modem-day society faces. Unification Thought is not a theory for a theory but the key to the solution to actual problems. Today, living in this age, we sometimes experience the savage side of man. It is just like especially when we see the terror attacks, religious and racial strife among men. The family, youth problem in our society has reached a serious point. As you all know, we have access to unimaginable erotic pictures just a few clicks away on the internet. Our young people are exposed to this kind of obscene internet environment and cyber space.

Today, many people try to solve the conflicts and violence in this world with force. The Iraq war is a good example of this rationale. But in the end this method must rely on the theory of force whether it is political, economic, and military might. Violence leads to even greater violence. Mutuality is the basic characteristic of violence. Thus violence only breeds to greater violence. In other words we cannot find the ultimate solution to our problems through violence.

The recent conflict between Islam and Christianity is like that of a fight between brothers. They both believe in Abraham as their ancestor of faith. Thus we can regard their struggle as the struggle between Abraham's two sons, Isaac and Ishmael. It is the only parents who can solve this struggle between brothers. Therefore the way for the peaceful resolution to these problems can be accomplished only through heart, love, and reconciliation. Isn't that so? It is the same with the ideological problems. We cannot solve the conflicts and struggles in this world without a philosophy based on heart and love. That is why the ultimate solution to these problems we face today can only be solved through philosophy of true love, namely the Head-wing Thought or Godism. This is the only alternative.

The Cold War has not ended especially in the Korean peninsula, and the nation is still at a face-off due to ideological conflict. That the Soviet Union, a communist state has collapsed does not necessarily mean that the substance of communism has gone away. Rev. Moon is the only one who can save communism and lead it towards the right path. Ultimately we must solve all the religious, racial problems, the problem of North and South division, ideological differences, the family problem, and religious conflicts with the Unification Thought which will become our strong weapon in solving the world problems. It is not until a new dawn approaches that the darkness of night dilutes. Thus when the light of God's thought and truth is revealed, the world of darkness will soon be dispersed.

I sincerely hope that this symposium does not end in the exchange of the opinions of various scholars, but by applying Unification Thought to your own area of specialization, a turning point for forming and reconstituting new sciences win be achieved. When this movement influences the scholarly world peace in the era of global era will come even more quickly. It is my hope that through this symposium, Unification Thought will be developed into a movement of sciences. In this sense, this symposium builds up a flow of new sciences with Unification Thought.

The papers to be presented this time are new in terms of their composition and creativity, and each bears the hard work of the presenters. I would like to request all of you to show keen interest in them and hold lively discussions on them. Moreover I truly hope at this moment, the meeting of the oriental and occidental scholars centering on Unification Thought becomes the corner stone of a peace movement. Thank you very much.

The Unity of Sciences and Unification Thought

Keynote Speech at the 17th International Symposium on Unification Thought

December 2, 2005, Keio Plaza Hotel, Tokyo, Japan

Distinguish scholars and scientists, Ladies and gentleman! I would like to thank all the participants for their academic interest and efforts in helping this symposium fruitful, especially those who have traveled distance and taken valuable time from their very busy schedules to attend the 17th International Symposium on Unification Thought.

Unification Thought is expected to play a key role in accomplishing the ideal of uniting sciences. In 2002 the Unification Thought Institute held the 14th International Symposium on Unification Thought in Tokyo under the theme of The Unity of Sciences and Unification Thought. After that, under the same theme the 15th Symposium was held in Moscow in 2003 and the 16th symposium in Sofia, Bulgaria in 2004. This time the 17th symposium is being held in Tokyo to continue the efforts.

Permit me to give a few remarks, as the President of UTI-Korea, in relation to the theme of this symposium, "The Unity of Sciences and Unification Thought." In the recent centuries, science has had an astonishing success in providing explanations of a wide range of natural phenomena. Such is this success that it would seem to be foolish and reactionary to resist the physicals claim that eventually all phenomena, including even the most subtle of human thoughts and actions, will be completely explained scientifically.

The unity of sciences is surely an ideal, which is present, consciously or unconsciously, in the heart of every scientist. The unity of sciences appeared as an ideal to strive for when physics and chemistry developed with immense rapidity in the 17th century. Newton already united the descriptions of two phenomena: the motion of objects falling to ground, with the revolution of the moon around the earth and the planets around the sun. The successes of the physical sciences raised the hope that ultimately the behavior of all objects could be described by laws which could be obtained through the study of material objects.

After the Newtonian success, another fundamental unity established by modern physics was that of Einstein's relativity theory. A moving body becomes heavier at a slower pace than a resting body of the same kind. Space and time are thus intimately connected to the point of mutual convertibility at a fixed rate, which is measured by the velocity of light in vacuum. This new unity is the inseparable field called space-time.

Subsequently, the emergence of field theories led to a unified treatment of electricity, magnetism, and light; quantum mechanics may be viewed as a formal way of unifying the corpuscular and wave-packet aspects of physical reality. That part of quantum mechanics is based on Schrodinger's equation, which is consistent with our knowledge of cosmic phenomena.

But, for microscopic physics, in other words, quantum mechanics, space-time coincidences do not exist. Nevertheless, the physical scientist still goes in search of the harmony, or theoretical consistency, between macroscopic cosmic physics and microscopic quantum physics. Quantum physicist Schrodinger, who is also one of the founders of molecular biology, searched for the problem of 'life' in relation to quantum mechanics. So, it can be concluded that a greater unity of sciences is demanded not only by the present state of physics, but also by our general conceptions of science.

Concerning the Movement for the Unity of Science, we know that scientists have tried to carry out the task to accomplish the ideal of science for a long time. To sum up the history of that movement, the first attempt to realize the ideal of the Unity of Sciences along with developing science was the movement of the School of Encyclopedia based on the philosophy of enlightenment in the 17th century of Europe. They compiled an encyclopedia on the basis of empirical philosophy of enlightenment and mechanical materialism, the predecessor of The Encyclopedia Britannica, and are enhancing its reputation even today.

The second attempt was the Unity of Science Movement centering on the Vienna Circle. In the 1930's, the Institute for the Unity of Science was set up under positivist auspices. Its slogan, "Unity of Science" was created by the living dynamo, O. Neurath, and was taken up by other members, in particular, R. Carnap, H. Reichenbach, C. G. Hempel, P. Frank, physicist P. Bridgman, biologist G. Wald, statistician R. von Mises, logician W. V. Quine, psychologist B. F. Skinner, young linguist, N. Chomsky, and various sympathizers of logical empiricism in England, Germany, Austria, Poland, France, and the United States. They began publication of International Encyclopedia of Unified Science, which laid the foundation for a new methodologically unified approach to cognitive assertion in general.

Their original thesis of the unity of sciences was based on the belief that all scientific concepts are definable in the strictest sense of language of sense data by terms belonging exclusively to the positivistic language, namely, the physical language. Carnap originally formulated the thesis of extension-ability as one of the mainstays of the unity of sciences.

Today the claims of Vienna Circle look rather naive. Most logical empiricists came to realize that most terms in scientific theories are neither definable by, nor reducible to, the so-called positivistic language. The Institute for the Unity of Science has recently dissolved. That its aims would have been achieved might well be doubted. The original positivist manifesto has been abandoned.

As I mentioned above, the two movements for the Unity of Sciences with the development of science did not play the leading role in realizing the unity of sciences. If I point out the causes of the failure of the movements for the unity of sciences from the standpoint of Unification Thought, it is that the philosophical foundation of the two movements was put on mechanical materialism and materialistic positivism.

Contrary to the inductive method of logical empiricism, Aristotle's attempt in the early period of Greece was also to construct a system of all sciences based on formal logic. Science in his view must rest on first principles, or necessary truths, certified as true by the skilled insight of the philosopher-scientist. From these premises, deductive logic can generate demonstrative science in which the more specific conclusions appear as necessary truths. But Aristotle's deductive logic could not become the universal principle to establish the foundation of metaphysics and science.

From the viewpoint of Unification Thought, the system of formal logic of Aristotle is literally formal system, which is a vacant logic system to tell nothing about the empirical essentialities and empirical science, which can become the basis of science. To establish an eternal and necessary foundation, as Aristotle mapped out, the deductive motive coming from the mind, sphere of sungsang, and the inductive motive of the level of experience, in other word, sphere of hyungsang should be under consideration. In Unification Thought, when these two motives, that is, deductive motive and inductive motive, are under consideration in the sense of subjective and objective, the basis of true science can be secured.

Unification Logic recognizes the a priori sphere of subjective conditions as foundation for the possibility of positive experiences, and as the control of all empirical conditions. Accordingly, from the viewpoint of Unification Logic, the foundation for proper science can begin neither empirical 'positivity' alone nor subjective a priori alone. It is Unification Logic, which is being promoted as an alternative to formal logic, which has been the basis of most academic endeavors.

Unification Thought intends to systematize the thought of Rev. Dr. Sun Myung Moon and present it in an appropriate order. Unification Thought is presented as a system that can offer a solution to the confusion affecting human and social sciences, the natural sciences, and the arts. The unity and harmony, which it brings to the various theories competing in these fields, must be based on the fundamental principles of the universe. Unification Thought aims to realize a world in which all humankind can serve God as one great family and bring about a peaceful world. This intent may provide insight into why Unification Thought is also referred to as Godism or Headwing Thought. Godism refers to the fact that this thought has God's truth and his love as its core.

Rev. Dr. Moon introduces God, using the metaphor "Scientist of Scientists and Artist of Artists." He emphasizes the importance that at the foundation of science, we should hold theistic presuppositions, on the basis of Godism and Unification Thought, because God creates all natural things and the world. Therefore, no science or scientific research can be justified unless it is based on God. That is why, spending an enormous expense, Rev. Dr. Moon held ICUS (International Conference on the Unity of Sciences), inviting intellectual elites around the world and advocated a quest of science centering on Absolute Value and a movement for the Unity of Sciences. Rev. Dr. Moon asserted at the 4th ICUS held in 1975 as follows:

Today, in all the fields people are prone to narrow their research to small and limited areas so that they tend to lose the overall purpose or the centrality of their subject..... In order to integrate all the specific fields of research, we are in need of a larger design or blueprint. In this way we may have a common ideal before us as we proceed to achieve this integration. The main purpose of this ICUS is for us to produce that blueprint.... Man is aware that he has life within him because his mind, which originates from a source of the highest dimension, is not limited to space and time. That source may be called the cosmic mind [true love or Shimjung] or the first cause of all beings. Man must be able to understand the centrality of absolute value [love] in the cosmos in order to give his life meaning.... We must not lose the very central point of the whole purpose: science is not for science itself but for the welfare of humanity.

I sincerely hope that by applying Unification Thought to each area of academic field, a turning point for forming and re-constructing new science will be reached. When this movement influences the scholarly world, peace in the era of global village will come even more quickly. I also hope that through this academic endeavor, Unification Thought will become a base for a movement for the unity of sciences. In this sense, Unification Thought-based movement will build a momentum for a new science as the centripetal point.

One unique aspect of this symposium is that Unificationist scholars from Korea, Japan and the United States are going to give presentation based on their research, submitted at ICUS in the past. They selected important papers from among many, and examined them from the perspective of Unification Thought. As a result, more than 9 volumes of booklets have been published in the series Selected ICUS Papers and Unification Thought.

May your intense discussions during this symposium be successful and most fruitful. I pray for God's protection and blessing to be with you and your families and your nations. Thank you.

THE CRISIS OF THE MODERN CIVILIZATION AND UNIFICATION THOUGHT

KEYNOTE SPEECH AT THE 18TH INTERNATIONAL SYMPOSIUM ON UNIFICATION THOUGHT

DECEMBER 1, 2006, KEIO PLAZA HOTEL, TOKYO, JAPAN

I am deeply honored and highly delighted to be able to speak at this 18th International Symposium on Unification Thought. In this keynote address, I would like to speak on the theme of 'The Crisis of the Modern Civilization and Unification Thought'

The reason why I place a special emphasis on the emergence of the world civilization here is that we are now faced with the reality where modern civilization, which has had overwhelming supremacy up to now, is failing to perform its proper function and on the verge of decline. In the course of its expansion throughout the world, the modern civilization has lost its dynamic vitality and is facing a crisis. The dynamism and creativity this civilization once possessed before it entered the modern age has now lost persuasion and started on a spiral of rapid downfall.Originally, modern western civilization was highly dynamic, had creative vitality, and played a magnificent role on the historical stage. It is true, however, that this civilization's encounter with numerous other civilizations around the world in the modern age has generated negative consequences in many respects. Now, I would like to point out a few flaws in the modern western civilization.

Through the two world wars, mankind, which had firmly believed in human 'reason', became startled at human brutality, and cast doubt on human 'reason' as the basis of modern civilization. Furthermore, modern industrialized society, characterized by mass production and mass consumption, has brought a serious distortion to human nature while providing living conveniences. The human nature created by modern materialistic civilization and competitive society is truly terrifying. The modern person every day has to wage a war of all against all (bellum omnium contra omnes) in the jungle of the survival of the fittest. What is ruling modern mass society is power and jungle ethics of the survival of the fittest. This is an outcome of dialectical philosophy, which understands the self and others in the relationship of conflict. This way, through the materialistic civilization, modern people have lost sight of their own dignity and worth; the value of human beings has drowned in the culture of materialism, and the loss of the original human value has precipitated human identity crises.

The crisis of modern civilization does not stop here. The structure of the modern industrialized society agrees more with utilitarianism, which pursues the greatest happiness for the greatest number, than with the stern moral philosophy of Kant. The inventions of the modern civilization such as automobiles, refrigerators, and laundry machines are bringing conveniences and comfort to modern people, and as these people see things from the standpoint of convenience and comfort, they cannot but be allured and captivated by utilitarianism and hedonism.

Moreover, such social crises in this industrialized society stems from a collapse of the basic and proper order of human relationship. Machine operations surpass human performance, and the overall industrial structure based on machines has humbled humans and has thoroughly degraded them into being mere parts of machines. The organic worldview, which understands human beings as the principle of life, has lost persuasion, and a mechanical worldview, which construes them in terms of organization in the framework of a machine, has gained supremacy in the modern society. Therefore, human relationships have ceased to be spiritual encounters, as we are isolated and alienated in the bloodless structure of organizations finding ourselves in the mechanical framework. In short, we have become slaves of the modern materialistic civilization.

Another point I would like to mention with regard to the problems of the modern civilization is the crisis of the confusion of values. In this age, view of values is overturning, and the meaning of life is distorted primarily because the relationship between God and human beings has been dismantled. A person who has lost sight of God can only see spiritual barrenness, existential emptiness, and the meaninglessness of life. Only where God is served at the center, can the true and ultimate meaning of human beings and the universe be restored, and the meaning of life and foundation for absolute value found.

Human beings serving as the standard of all values can only give rise to thoroughly relative values, as exemplified in the worldview of the sophists of ancient Greece. Then, as we see in the modern society, materialism and worldly greed may be accepted as the standard of value judgments, and atheism and hedonism may proclaim their legitimacy and gain power.

To enumerate some of the philosophies that have had the greatest impact on modern society, we can list Karl Marx's dialectical materialism, Darwin's evolutionism, positivism of Conte, and so forth. These atheistic philosophies have caused a serious confusion of values for the moderns as they elevated materialistic values above spiritual ones while distorting and laying aside absolute values centered on God and human dignity.

Earlier, I have pointed out a few crises of modern civilization. These crises have become a general situation of the age rather than an isolated or temporary phenomenon. Such symptoms of modern civilization are offshoots of the limitations inherent in it rather than caused by its confrontation with other civilizations. During its progress through antiquity, the Middle Age, and modernity, western civilization maintained vitality with its own unity and creative energy; however, when it clashed with other civilizations in modern times, its limitations became exposed. Modern western civilization has become too weak and old to be able to lead a world civilization. Its expansion throughout the world by means of its economic power and imperialism has ironically created an opportunity whereby its inner sickness surfaced into a full-blown disease.

Many people have foretold a shift of the center of civilization from the West to the East. Western civilization, which began with the magnificent Mediterranean civilization of Greece and Rome, developed through the age of the Atlantic civilization and now has reached its limit and is at a standstill. Judging from the westward movement of civilization, Western civilization, after the age of the Atlantic, will be succeeded by the Pacific civilization, in which Western civilization will blossom and yield fruit. The creativity and energy to lead the world civilization lies hidden in the Pacific civilization, which is now about to emerge. At this point, I would like to suggest a few characteristics of the philosophy that will lead the world civilization.

First, the philosophy to lead the world civilization must have a firm moral root. The progress of society and culture cannot be measured in terms of technological advance or material superiority; rather, the maturity of a society and culture should be estimated by the standard of morality. When the leaders of a society lose their moral persuasion, the society is dissolved, and just the same pattern applies to the philosophy to lead a world civilization. We can see this in the ethical philosophy of Socrates, which delivered a fatal blow to the relativist and skeptical philosophy of sophists, which had thrived parasitically on the corrupt democracy of Athens. Likewise, a true philosophy should be able to speak for the conscience of the age with moral persuasion in times of crisis.

Second, the philosophy to lead a world civilization must be able to embrace both the civilizations of the East and West and make a new creation out of their integration. It should bring harmony to the materialistic civilization of the West and spiritual civilization of the East. Also, it should be able to harmonize the technical culture based on analytic and logical thinking of the West and ethical culture based on the intuitive thinking of the East. While not ignoring the principle of efficiency, the leading philosophy should lay stress on ethics of the heart.

Third, the philosophy to lead the world civilization must discover God, lost in modern civilization, and adopt absolute values centered on God as its foundation. Out of an excessive emphasis on human reason, Western civilization has dethroned God, and replaced Him with a humanist philosophy, even promulgating atheism and materialism. Human beings detached from God have slipped into self-worship, and their drive for conquering the environment has ravaged the ecosystem and even endangered the very foundation of their life. Freedom apart from God, however, cannot be true freedom.

It is urgent that we teach a new God-centered system of values from the university level. Plato's Academy, the first university in history, had a vast system of academic disciplines, performing ceremonies for the God-Muses, teaching mathematics, philosophy, natural science, and so forth for the purpose of the catharsis of the human soul. The dialectics, which they adopted as academic methodology, was none other than a process resembling God-supreme universal, that is the 'Idea of Good'. The spirit of the Academy was that all the studies and education find mutual connection in reliance on God. Today's school education, however, has lost the original Academism, being degraded into a field of transmitting knowledge and technological training. It is the first condition of the philosophy leading the world to establish an absolute and God-centered value system.

I see that the philosophy that is best equipped to lead the world civilization of the future is the Unification Thought advocated by the Rev. Dr. Sun Myung Moon. The Unification Thought is a new system of philosophy to bring proper order to the relationship between God, human beings, and nature. In this, we come to discover the true position of God, the meaning of human existence, the social meaning of human relationship, and harmonious order of human beings and nature.
Unification Thought, as well as overcoming the limitation of the Western philosophies, re-establishes the importance and meaning of the Eastern philosophies. There, we come to find the sources of Confucianism, Buddhism and Christianity, re-interpreted in light of Unification Thought.

My sincere hope is that this symposium will not end in the exchange of the opinions of various scholars, but create a turning point to construct a new science by applying Unification Thought to your own area of specialization. Having a new perspective in science is of paramount importance. When this movement influences the scholarly world, peace in the era of global village will come even more quickly.

It is my hope that through this symposium, Unification Thought will be developed into a movement of science. The centripetal point for shaping science is Unification Thought, and we need to tap the possibility of unifying the various thoughts with Unification Thought as the axis. A special cultural characteristic of the nineteenth and twentieth centuries,_which is the worthiest of close attention, is related to the formation of a school of science. Unification Thought has the specialty of not being able to be prescribed with the mundane features of school since it unifies all sciences with the absolute value system centering on God. In this sense, this symposium is expected to build up the flow of a new science with Unification Thought as the centripetal point. Hence I think this should be a gathering by which the movement of science with a new viewpoint quickens.

Now we must overcome the obstacles to the formation of an world community such as national selfishness, a drive for national supremacy, and nationalistic grudges, uniting in the spirit of cooperation and living for the sake of others. This way, I sincerely hope that in the new millennium God's Will be realized on earth. Thank you.

UNITY OF SCIENCES UNIFICATION THOUGHT

KEYNOTE SPEECH AT THE 19TH INTERNATIONAL SYMPOSIUM ON UNIFICATION THOUGHT

DECEMBER 1, 2007, KEIO PLAZA HOTEL, TOKYO, JAPAN

I.   Unity of Science God

'Unity of science' is an ideal as old as science itself. Unity implies consistency. Inconsistency of statements, even if pertaining to different branches of science, is incompatible with science. Many other reasons, including the urge for systematization, coherence, simplicity, and some purely aesthetic consideration, motivated this quest for unity. It also seems to have been early recognized that unifications are apt to increase our knowledge.

The collective effort of a unified theory will be more productive than the sum of what the individual theories could accomplish working separately. In any case the unified theory is logically and semantically stronger than each partial theory, that means the nomological patterns get more tight and a surplus meaning is generated which cannot be grasped in separation. In physics, unification also means to go up to a domain of higher energy. The gauge theory description reveals that the world becomes much more simple at extreme energies, when symmetries which are broken within our cold universe are restored.

Moreover, when mathematics and the motion of bodies were brought together; when atomic theory and the periodic table of elements were brought together; when it was discovered that the functioning of the living cell was its molecular structure; and when thermodynamics and mechanics united through Boltzmann's statistical theories, science became certainly progressive research program in the direction of the unity of science. It was

Einstein's conviction and it is today's theorists' too, that nature itself bears a deeper unity.

In case of the Unity of Science Movement of the neopositivists and its extremist form of physicalism, unity of science is without doubt an ideal of science that must be realized. Here unity of science is not something that exists already but the anticipation of an ultimate result though a cooperative study of the logical languages of sciences that should be in accordance with the trend of modern science.

From the viewpoint of Unification Thought, completely accepting normative issues on the ideal for unity of science stated above, and moreover, how to establish the absolute value or Godism on the all scientific foundations are raised. Because, sciences is a process to investigate universal truth, and the universal truth is only revealed accompany with absolute value and norm. Just like Unification Thought, for Plato, issues of knowledge and the norm of practical life are not exclusive each other, since systems of truth and value are related to a same thing. And on the center of them, under the mode of absolute value, God's existence is assumed. In the western tradition, physics of Platon, Aristoteles, and Kant, as well as those of Kepler, Newton, and Einstein show that they are closely related to the existence of God. The prospectus of the 19th International Symposium on Unification Thought is as follow;

"The ultimate Being, exists as the fundamental cause of all existing beings, no matter how differently it is called (Allah, Jehovah, God, etc.), is one and the same. Thus, it is logical that if the attributes of the ultimate Being are clarified and if the method of the design and creation of the universe are sufficiently and logically (even scientifically) clarified, the major teachings of all the religions can be fundamentally and harmoniously integrated and unity can be promoted. Godism offers a logical theory of God which is refreshingly open to academic and scientific examination. It maintains that God, as the origin of the universe, is, above all, a Being of Shimjung (Heart and Love). As such, it advocates the realistic idea of building and developing a Culture of Heart, as a means of realizing a world based on heart and love.

Unification Thought, in presenting universal principles, offers a positive challenge to the sciences to undertake a common approach to truth through uniting their academic disciplines in a more integrated manner. This symposium attempts to inherit the achievements of the International Conferences on the Unity of the Sciences (ICUS), which, based on the vision and proposal of the Rev. Dr. Sun Myung Moon, were held many times since 1972, and it seeks to expand and enhance the achievements of ICUS, from the perspective of Unification Thought."

In early period of Greek, the task of the wise man, the philosopher, is to discover universal principles and provide a basis for episteme, for knowledge. Such knowledge would have about it the qualities of eternity and necessity that would link it to the Divine.

Plato had already known that common language cannot be used consistently for the description of the unity of the universe. Plato mythicized, then, an account of how the Demiurge, i.e, Creator/God, fashioned things out of the receptacle (chora) using the Forms as patterns. The World Soul is produced by the Demiurge and is the energizing activity in the receptacle, producing what to us appears to be substance or solid matter though in reality is only qualities caused by the rearrangement of geometric surfaces.

Here Plato was forced to assumed that all things must be ordered by spirit, that the cosmos is the activity of the World Soul in the receptacle. In short, Plato's point is that all things which is something perceived is to be appear and perish, and the permanent factor is rather the receptacle as mathematical pattern, and this has been connected to Idea and Form which is fashioned by God. Modern quantum physics can be seen like that succession of Plato. Especially in Heisenberg and Einstein. The following passage by Patrick A. Heelan, one of the scholars of the interpretation of the modern physics, confirms this fact most eloquently.

"Heisenberg's view was more complex: from Pythagoras, Plato and Democritus came the impulse to explain immediate and direct perceptual qualities by the geometrical and other mathematical properties of fundamental non-perceptual entities, atoms and the like: the history of modern science is the story of the successful growth of a mathematical explanation of nature."

The big leaders of quantum physics like Bohr, Heisenberg, Schrödinger, and Pauli knew very well that the combination of mathematics and theology which began with Pythagoras characterized religious philosophy in Greek, remained in the vein of Plato, and down to Kant.

II.   Universal Principle in Nature

The question to be addressed by our topic is: "what are the characteristics of universal principle". It is part of an overall inquiry into the conditions for a "unity of science". But before we can ask about the characteristics of universal principles, it is necessary to give some attention to the evident presupposition of this question, namely that there are such principles.

Indeed, when Thales, the first Milesian monists, declared "All is water," he expressed in what is generally regarded at the first scientific statement in the material unity of nature. When Anaximander introduced the notion of Apeiron, or boundlessness, and when Anaximens formulated his theory of the condensation and rarefaction of air, they prepared the idea that one regulative principle governs the nature.

For it was the Pythagoreans who with their mathematization of physics diverted the search for a nomological unity of physical reality. Pythagoras is said to have been impressed by the manifest presence in the world of mathematical harmonies. He postulated in consequence, that the structures of the perceived world are at bottom mathematical structures. It would follow that a science of nature would reduce to mathematics. Pythagoras would equate mathematics and physics, mathematical entities and physical entities. Aristotle argued that this move could not but lead to incoherence. The universal principles of such a science would be entirely emptied of physical content and would be judged on their formal unifying power only.

In view of Unification Thought mathematical principles are the ultimate cause for the mathematical phenomena inherent in the natural world. All numbers, mathematical values and formulas, which lie in mathematical phenomena, come ultimately from the mathematical principles in the Inner Hyungsang. In this respect, Pythagoras showed that numbers which is rooted in all things.

The Democritean atomists move is the basis of all forms of reductionism. In this view, the unity of science must be attainable because all sciences ultimately reduce to the most general science employing sensible data, i.e. to physics. But it does not work that "primary" concepts are never really primary enough, that is, they are never definitive, given once for all.

Plato's Form is the basic principle of knowing, Form is also one of the basic principles of being. Of itself it would not be sufficient to constitute the world we know, since it accounts for sameness but not for individuality. Nevertheless, in this perspective, the status of the principles as universal in scope presented no problem, since the most basic principles, the Forms, were themselves by definition universals, capable of indefinitely repeated instantiation.

Aristotle made perhaps the boldest suggestion of all. He provide a tight deductive and conceptual unity of science, since all conclusions/theorems have to be implicit in the premises from which the process begin. This puts a lot of weight on the process of intuition/induction by means of which these principles are grasped and warranted in the first place, i.e. premise. Following Popper's view, one may cast science in deductive form, but the "principles" on which it will rest will be tentative, always open to amendment. So, principles in deduction can't be universal principles.

The most general universal principle occurring in nature, in view of Unification Thought, is give and take action. The law of give and take action states the dynamic relationship of subject to object and vice versa. Because of this law we observe an order, subordination and harmony in nature; this law also accounts for maintenance of the identity of objects, that they are what they are. This self-identity subsequently determines their relationship to other beings.

In view of Unification Thought, laws or principles in the Inner Hyung-sang are the original laws at the root of natural laws and norms. In other words, it is in and through the numerous natural laws of nature and norms of human life that these original laws find their expression. By laws of nature is understood, in Unification Thought, a constant and repeated activity of beings. The general concept of the natural laws is applicable to every being, as can be concluded from their definition. The formulation and establishment of such laws is necessary in order to explain beings in a scientific and reasonable way and not in a mysterious or mythical one.

III. Logical Resemblance between Physics-psyche Duality in Quantum Theory Unification Thought

The new concept of a field also has been associated with that other major force in the large-scale world, the force of gravity. Wherever there is a massive body, there will also be a gravitational field, and this field will manifest itself as the curvature of the space surrounding that body. In Einstein's theory, then, matter cannot be separated from its field of gravity, and the field of gravity cannot be separated from the curved space. Matter and space are thus seen to be inseparable and interdependent parts of a single whole.

The striking new feature of quantum electrodynamics arises from the combination of two concepts; that of the electromagnetic field and that of photons as the particle manifestations of electromagnetic waves. Since photons are also electromagnetic waves, and since these waves are vibrating fields, the photons must be manifestations of electromagnetic fields. Hence the concept of a 'quantum field', that is, of a field which can take the form of quanta, or particles. In the words of Albert Einstein:

"From the relativity theory we know that matter represents vast stores of energy and that energy represents matter. We cannot, in this way, distinguish qualitatively between matter and field, since the distinction between mass and field, since the distinction between mass and energy is not a qualitative one. By far the greatest part of energy is concentrated in matter; but the field surrounding the particle also represents energy, thought in an incomparably smaller quantity. We could therefore say: Matter is where the concentration of energy is great, field where the concentration of energy is small."

Quantum theory has shown that particles are not isolated graina of matter, but are probability patterns, interconnections in an inseparable cosmic web. The particles of the subatomic world are not only active in the sense of moving around very fast; they themselves are processes. The existence of matter and its activity cannot be separated. They are but different aspects of the same space-time reality.

The new concepts of interrelation between a material object and its environment is not only a basic element of quantum field theory, but also a basic element of the Eastern world view. The intuitive world view of Eastern mystics, of course, cannot be identified with the quantum field of the modern physics. Nevertheless, there is no doubt that the intuitive world view behind the modern physicist's interpretation of the subatomic world, in terms of the quantum field, is closely paralleled by that of the Eastern mystic who interprets his or her experience of the world in terms of an ultimate underlying reality.

Subsequent to the emergence of the field concept, physicists have attempted to unify the various fields into a single fundamental field which would incorporate all physical phenomena. Einstein, in particular, spent the last years of his life searching for such a unified field. Also, Pauli continues:

"The general problem of the relation between psyche and physics, between the inner and the outer, can, however, hardly be said to have been solved by the concept of "psychophysical parallelism" which was advanced in the last century. Yet modern science may have brought us closer to a more satisfying conception of this relationship by setting up, within the field of physics, the concept of complementarity. It would be most satisfactory of all if physics and psyche could be seen as complementary aspects of the same reality."

We now understand that there exists the field of non-matter. The quantum mechanical interaction in the field of seems to be utterly non-matter and their interactions are explained by the principle of equation of quantum mechanics. These equations are not telling about mass in motion. In that way the paradox of the field shows more mysterious aspects than that of the magnetic or gravitational field. The electromagnetic wave, such as visible light and radio waves, exhibits two inherent properties, namely, a particle (photon), which is a packet of energy, and a wave, which propagates in vacuum. Here the concept of "field" was introduced to describe the " either" in which the electromagnetic wave propagate. According to Einstein's formula, $E=mc^2$ the mass is nothing but an aggregate of energy.

There exists a certain aspect of the world which cannot be explained by the mechanism of causation. The contents of the experiences of the modern physicists are beyond the realm of the classical concept of space and time. This experience can only be understood by the new concepts of space, time, and matter.

Heisenberg argued that the nature of Bohr's view of wave-particle complementarity is highly similar to that of Descartes, mind-body dualism, and suggests that even though the contents of consciousness have the nature of the non-material, they are profoundly connected to the nature of the physical brain. This also contributed Heisenberg's theory. The reaction quantum theory presented shows the perfect unification of the positivistic experimental observation with the a priori condition. Accordingly, in this sense it is suggested that there must be a complimentarity between empiricism and idealism and the physical and the mental.

Pauli emphasized that Greek philosophy and science are deeply influenced by mathematics and astronomy, keeping the mysticism of Babylonia, and he represented as evidence that the soul exists in everyone and oneness that's God is being expressed everywhere, as that number exists everywhere in the Pythagorean school.

Even the atom we can know only mathematically, not sensually, which keeps the mutual dependency between the concrete and the abstract and between the transcendental and the experienced which can't be excluded. In a word, today's modern science needs the philosophy of unification, including the experience of rational understanding and mystical unity.

The notions of form and matter, or spiritual and physical reality, as conceived by Western philosophy, have led to several contradictions. The Unification Thought notions of Original Sungsang and Original Hyungsang offer a solution to that crucial issue in that they are introduced as two different manifestations of one identical fundamental element. This notion in reality exactly identical with the mode of psyche and physics in quantum theory. At this point, we will consider whether sungsang and hyungsang are essentially homogenous or heterogeneous, which will lead us to discuss the issue of monism vs. dualism and other ontological questions. For Unification Thought, the Original Sungsang and the Original Hyungsang are two manifestations of one homogeneous core element. Energy and the mind do not exist as distinct entities; they are essentially one. When considered from the perspective of ontology, this viewpoint can be called "Unification Theory" or "Theory of Oneness."

So, sungsang consists primarily of mental elements, but there is some element of energy in it as well. In Sungsang, the mental element is predominant over the element of energy. Likewise, Hyungsang is made of energy, but there is some mental element included in it. This can be understood from the following example. In our daily life, we can move the muscles of our hands and legs by our thinking: Our thought stimulates our nerves and moves our muscles. This means that our mind has the same kind of energy as the physical electrical energy. So the logical structure of "Unification Theory" is similar to the mode of psyche-physics duality in quantum theory.

THE UNITY OF SCIENCES AND UNIFICATION
THOUGHT

TOWARDS EXPLORING UNIFICATION THOUGHT ACADEMIC DISCIPLINES

KEYNOTE SPEECH AT THE 20TH INTERNATIONAL SYMPOSIUM ON UNIFICATION THOUGHT

NOVEMBER 29, 2008, KEIO PLAZA HOTEL, TOKYO, JAPAN

I would like to express to all the distinguished scholars and observers my heartfelt welcome and gratitude for participating in the 20th international Symposium on Unification Thought.

In expressing the flow of the history of human thoughts straightfor-wardly, a colossal struggle of the conceptions of God lies at its basis. It would not be an exaggeration to say that human history is a history of the struggle between theism and atheism, and the conflict of the conceptions of God where different views of God are vying to emerge as the sole winner.

The basis of the first ever university founded by Plato in human history, the Academia, and the teachings of the Lyceum school established by his student, Aristotle, was also theology. Plato was the first person to use the term "Theology," and at the utmost zenith of the teleological view of the world put forward by Aristotle was placed the Unmoved Mover, that is, God. However, the God of Aristotle was the eidos of eidos, a purely mental Being, and this conception of God passed through the Scholar Philosophy and came to take root as the basis for the conception of God in the orthodox theology

of Christianity. As you well know, at the center of the theological controversy of the universal dispute that stirred up the entire Medieval Age was
Aristotle's view of God, and it is common knowledge that this view was the
measure of orthodoxy or heresy in the history of Catholicism.

Not to mention modern times, modern civilization can, in a way, be said
to be the endless fierce struggle to escape from the influence of medieval
civilization. This is because the atheistic ideologies that lie at the background
of the modern times and the modern scientific civilization were begun from
the challenge to the conception of God in Christian orthodoxy which was
passed down to us from the Medieval Age.

As we entered the present age, Marx and Darwin, Freud, scientific materialism and the like challenged in the name of science the Christian conception of God, and the theology of God's death influenced by Nietzsche's ideology and logical positivism that forms the mainstream of modern philosophy were also in dispute with this Christian view of God. Modern civilization can actually be said to have originated from the battlefield of fierce and
bitter struggles of views of God, which vied with the God of Christian orthodoxy in one big fight to decide the winner. If the Christian orthodox conception of God had been right, the atheistic challenge of the modern civilization would have lost meaning and come out the loser. However, the state of
things took a different turn, and it looks almost as if these atheistic challenges are in command of the situation. In the biological sphere, Darwin's
evolutionary theory is pretending to be the scientific truth and it is actually
studied in all educational fields around the world, and atheistic biology is
being nurtured. Strauss, the Hegelian leftist, asserts that Darwin's evolutionary theory, and not God, is the superior theory which explains about the
world more reasonably. He predicts that if Darwin's evolutionary theory
were to become the basis of ethics instead of Jesus Christ, then in the place
of ethics of the grace of God, compassion, love and so on, the power of the
strong demonstrated in the "war on all against all" and "survival fitness" will
become the new form of ethics.

Darwin, by substituting God's "Creation" with "Natural Selection", explains that even without God's creation such biological species like monkeys
evolve into human beings through natural selection. What I would like to
point out here is that I am not trying to appeal to the human feeling of anger
through this ludicrous argument that monkeys evolved into human beings.
Instead, the question is, where can we find the God who is excluded from
biology and life phenomena through atheism that is the doctrine of evolutionary theory and the atheistic ideology and mercilessness which reduces
the liveliness of life and the mystery of creation into cold materialistic biology and mechanism?

In the background of scientific materialism are two premises that scientific knowledge is the most believable and verifiable knowledge and that
matter is the most basic actuality in the world. Scientific materialists assert

that science is the only objective and universal knowledge which is at the same time also progressive. In scientific materialism, physical phenomena, natural phenomena, and even life phenomena are explained based on mechanism, and all fields of science must ultimately be reduced to materialism. Therefore, in scientific materialism there is no room whatsoever for God to intervene. Scientists can, without the supporting hypothesis of God, successfully establish theses in various fields of experimental science including physics, chemistry and biology, and they emphasize that, through materialism alone, with God excluded, it is possible to make successful scientific researches. Carl Sagan, the astrophysicist, asserted that only scientific methods can be universally implemented in the eternal universe and that faith in God run counter to finality in science and that the subject of human reverence should be nature and not God.

On top of all this God, excluded from the field of science, is losing His footing even in the field of theology. It is a well known fact that in the background of the Eastern Orthodox Church lays the philosophy of Plato, in the background of the Catholic Church lies the philosophy of Aristotle, and in the background of the Protestant Church the philosophy of Kant provides its basic ideas. However, Christian orthodox theology is based on Catholicism, formed through the theological disputes of the Medieval Age, so what historically determined the basis for this theology is Aristotle's philosophy and conception of God. Since Nietzsche declared the death of God, the theology of the death of God which emerged within the field of theology reveals its characteristics by expressing resistance and rejection towards the Christian orthodox conception of the mental God. Theologians of the death of God declare that Aristotle's conception of the mental God has no relationship with Christianity, which is a Biblical religion, and pose a question in the form of "What on earth is the relationship between Athens and Jerusalem?"

They assert that Aristotle's philosophic conception of God should be excluded from Biblical religion and that a new conception of God should be established in the tradition of Biblical religion. They have substituted their conception of God with Christology. This challenge shows an aspect of the very fierce struggle in the history of the conflict of conceptions of God. Be that as it may, could we define Jesus Christ put forward by the theology of the death of God as the God revealed in the Bible?

Of course not! If we were to understand God Himself to be the historical Jesus, the God of creation who appears in the traditional history of the Old Testament and the transcendent God who has worked His providence through the redemptive history of Israel become meaningless. I cannot delve deeper into this problem at this time, but to speak conclusively, the first error committed by Christian orthodoxy is the accepting of Aristotle's conception of mental God as the Christian orthodox conception of God, and its second error is the doctrine of combining God the Father and Jesus the Son through the thought of Trinity. The Bible clearly denotes the ontological difference between the Holy Father and the Holy Son, thereby showing that God the

Creator and the historical Jesus are not the same. From this historical error stemmed Hegelian philosophy based on the wrong premise; and such theology of absurd logics like secular theology and the theology of the death of God came to be.

The scientific civilization of the modern times presented scientific materialism and atheistic conception of God in opposition to the Christian conception of God. The reason I have named atheism as a form of the conception of God is that the atheistic basis of modern ideology forms a very strong tendency, and it has not merely stopped at denying "theism" but has gone on to exhibit a reckless religious inclination to construct a militant atheistic paradigm. Speaking figuratively, the scientific civilization of the modern times could be said to be assuming an aspect of a struggle for supremacy of a kind, to abdicate and exile God and to undergo a process of politic-ideological revolution and establish an atheistic regime.

The science of today, which boasted of universal and objective knowledge under the banner of "value-neutral science," has, contrary to its original intent, revealed a value-oriented propensity. In contrast to general knowledge, in spite of the fact that science is a process of researching theories to explain objective facts, scientists carry out their scientific researches based on certain premises. This premise and prejudice is none other than materialism and atheism. However, materialism and atheism are value systems of a kind that deviates from the field of science where researches are made on objective facts. The scientists of today knowingly or unknowingly carry out scientific researches premised on value systems of materialism and atheism, and this is the real state of affairs of today's' science and the irony of it, for science emphasizes value-neutrality.

Modern people, and modern scientists in particular, accept the science-materialism paradigm and the science-atheism paradigm unconsciously, and are comfortably indulged in their pastimes in the garden prepared by science. If the science-atheism paradigm can be established for the scientists, conversely the science-theism paradigm could also be established, but scientists do their utmost to disregard this fact. The more science excludes God and rejects the science-theism paradigm, the more it is recognized as the most scientific of all sciences, and the more a science declares atheism and asserts the science-materialism paradigm, the more it is acknowledged as being brilliant; such is the era of atheism we are living in now. Now, in the garden of science the science-atheism paradigm is reigning as the orthodox and the fittest, and the science-theism paradigm has been reduced to heresy.

Originally, most modern scientists like Galileo, Kepler and Newton were believers of the science-theism paradigm. Newton's study of physics was begun with the motive to give proof of God in the face of the challenge of atheists. Modern sciences like dynamics; physics and astronomy were inspired by the work of searching for God inherent in cosmic laws. However, the situation of science in modern society in which atheism is rampant has come to assume an aspect different from the beginning. Modern scientists, who have excluded God from the field of science, are now trying to exile not only God Himself but also the mind (consciousness) from that field. By constructing an elaborately prepared science-materialism paradigm, they are trying to expel God and the mind from the field of science. And they are showing clear signs of wishing to defend pure materialism and remain as materialists.

In tune with the age of atheism, most scientists are exhibiting a tendency to follow the general trend and get on the bandwagon of science-atheism paradigm. Incidentally, the pair of science-atheism and science-materialism paradigms have the advantage of appealing to simplicity. Science asserts that this pair of atheism and materialism paradigms can establish scientific theories with only matter as their parameter without having to import such parameters as God or the mind. Laplace, who disclosed the nebular hypothesis more generally known as Kant-Laplace hypothesis, insisted, "There is no need to postulate God in science". They assert that in science, scientific materialism excluding God and the mind is more suitable to the ideal of science which prefers simple theories. They also boast that it is in context with the aim of science which says that the simpler a theory the better. In a way this shows how the idea is established that the more God and the mind is excluded the more excellent a scientist becomes.

However, the situation is not that simple. The science of today is cornered to such a degree that, without taking into account God and the mind as parameters, it cannot break from its current limited state, and it cannot move forward any more. In the background of the quantum theory in physics, the big bang theory, the chaos theory, and the recent theories in brain physiology lies without a doubt the idea of the Designer of the universe and the theistic paradigm related to all living things in the entire universe.

Theism and atheism are value systems. And in science, theism and atheism cannot both be true. One of them must be true and the other must be false. In a stricter sense, the "value-neutral" ideal of science is an illusion and a scientific myth. Science is based on value and so it is value-dependent, and therefore it cannot be established separately from value.

Kant's dichotomy of fact and value, in which he considered fact and value separately, has become unconvincing, and it has been proved that blind fact apart from value cannot be established. For example, when we see a flower, our acknowledging the fact that "It is a flower" and our acknowledging its value as we see the flower by thinking "It is beautiful" are not separate, but instead they are interlocked so that we see the "Beauty" and the "Flower" at the same time and come to smile. We do not see the "Flower (fact)" and "Beauty (value)" separately. As we can see from this simple everyday experience, the value-dependency of a fact is also implemented in science based on fact. In this way, all facts are value-dependent and so science based on fact is also value-oriented. In this regard, Plato was right and Kant wrong. Science, whether it is theism or atheism, or materialism or mind-body dualism, must be established on any one value system. It cannot receive or reject both.

Here, we cannot but pay attention to the insight of Rev. Dr. Sun Myung Moon, who has intended to build the edifice that is science on the foundation of absolute values. And we should remember his tears and sweat as he devoted all his energies into creating a new civilization with God at the center, by advocating the International Conference on the Unity of the Sciences and investing a prodigious budget. The absolute values advocated by Rev. Moon are the values centered on God, and the conference signifies constructing a world of science with God at the center. This is because the absolute values of Rev. Moon call to mind 'the idea of goodness', which is the θeos (God) of Plato, and to Plato's θeos corresponds with the standard of values, which are the absolute values.

As can be seen, the International Conference on the Unity of the Sciences is a necessary enterprise requisite to restoring the God who has been exiled in the atheistic age and constructing a civilization centered on God. In order to know the significance of the International Conference on the Unity of the Sciences, we first need to understand Rev. Moon's viewpoint in interpreting modern scientific civilization, and further, we should understand science and absolute values, the definition of God, and the need of the age to hold the Conference immediately. However, what is clear to us is that the International Conference is related to the question of how God and the absolute values lost by the modern scientific civilization can be recovered and restored. Then who is the God lost by the modern civilization, and where is He?

The most ultimate and realistically important subject in modern civilization is the matter of God, and establishing the right conception of God is more urgent and fundamental than anything else. Unification Thought is related to the work of establishing the right conception of God. That is why some people wonder whether Unification Thought is a theology or not.

Philosophy and theology have this point in common that they both regard God as the object of their researches, but they differ in that the study in philosophy is based on reason whereas the study in theology is based on faith. In this way, theology pursues revelation through faith whereas philosophy pursues the truth through reason. In this regard, since Unification Thought is not premised on faith, it cannot be said to be a revealed theology based on revelation. However, as long as the subject dealt with in theology is studied within the scope of reason, it can become the subject matter for the research of Unification Thought. Unification Thought gives attention to the universal ideology and teachings given by Rev. Dr. Sun Myung Moon in his speeches, and tries to systematize the truth disclosed through our reason.

Scholars of the world express surprise at the fact that Unification Thought deals not only with theology or philosophy but also all sciences in general including humanities, social science, natural science and artistic fields. Moreover, they are even more amazed to learn that it presents solutions in all of the fields. Apart from whether Unification Thought is right or wrong, people express incredulity at the fact that it deals extensively with all fields of academic disciplines.

Then is Unification Thought a philosophy, a faith system, an ideology or an academic theory? The more the study and discussion on Unification Thought is carried out by world scholars, the more the question of the nature of Unification Thought comes up. Unification Thought's understanding of God does not try to prove the existence of God epistemologically or ethically, but has its aim in knowing clearly the true form of God. Rev. Dr. Sun Myung Moon, after studying in detail the all things in the universe and the central contexts of the Bible, has come to the conclusion that God must be a God of dual characteristics, and after implementing this to nature and the universe and history and the context of the Bible respectively, he has been confirmed in his belief that his conclusion was right. This method used in Unification Thought's understanding of God can be said to fall under the methodology of academic theory, similar to the hypothetical deductive method used in science.

Scientific theories commonly used in science, such as Newton's theory of dynamics and Einstein's theory of relativity, basically have a hypothetical deductive nature. The scientific theories of today are not derived from a process of induction based on experience, but instead it has come to be widely accepted that these theories are derived in a hypothetical deductive method through the scholar's intuition and imagination. Scientific theories are made based, not on "experience" and "induction," but on "reason" and "deduction."

As can be seen, scientific theories fundamentally acquire a hypothetical deductive nature. In this regard, religious truth and scientific truth shares a common character in the psychological aspect. The life of Abraham, spent in pilgrimage to some place or other in accordance with revelations or religious intuition, or the life of scientists spent in presenting some theory or other in accordance with their scientific intuition, are both based on intuition, reasoning, imagination, and belief that it will be so in the future.

Unification Thought's methodology in regard to the Theory of Original Image is, similar to scientific methodology, of a hypothetical deductive nature. Looking at it from the aspect of methodology, Unification Thought's Theory of Original Image has a hypothetical deductive system. However, same as in the case of scientific theories, for the theoretical system of the Theory of Original Image learned through intuitive knowledge to not only remain simply as a hypothetical system but to remain as the truth, it has to be verified through experiential facts. From this viewpoint, Unification Thought's Theory of Original Image gives evidence to unmovable facts whether it is implemented in the natural universe, human history or any part of the Bible, and so it does not end as being only one of several hypotheses but instead can be said to be a theory and a truth system with a firm foundation. In Unification Thought, the many problems that have to do with human beings, all things in the universe, society and the world, art, values, culture, history, education, science and so on, are all theories which can be deduced from the Theory of Original Image in relation to God. That is why the Theory of Original Image becomes the basis and the standard for all theories. And this Theory of Original Image is not merely a hypothetical system but it has been confirmed and verified through experiential facts in all fields of learning, and so it can be said to be an indisputable academic theoretical system and truth system.

We were able to observe the true character of Unification Thought in the First International Unification Thought Specialty Seminar held at the Hotel Nakadaya, Atami City, Shizuoka Prefecture, Japan for four days starting from September 4th of this year. 25 scholars majoring in physics from various parts of the world heatedly discussed the subject of "God, Spirit, and Physics" as seen by Unification Thought and excellent results were attained. Though there is not enough space to make mention of this in detail, through the International Specialty Conference we reached the conclusion that modern physics can prove the theoretical legitimacy of Unification Thought, and everyone voted to establish physics based on Unification Thought. I believe it is not mere coincidence that not theology or philosophy but physics, which is one of the pure sciences, harmonizes well with the theoretical system of Unification Thought. If Unification Thought is a true academic discipline, its theoretical superiority should be proved not only in the fields of theology or philosophy but also in the fields of natural science beginning with physics, social science, humanities, and art.

In conclusion, Unification Thought is not a science of a particular field like theology or philosophy, but it is the basic science of all sciences with its academic methodology and system, and it is a true academic discipline on the basis of which can be established all sciences. Moreover, in the secular civilization which sneers that God is dead and that all are well off without God, Unification Thought is the ideology of Godism which has declared the existence of God and is trying to find Him and construct a theistic set of values and a theistic paradigm in all fields of academic disciplines. And it is also the ideology of Rev. Sun Myung Moon which aims to systematize all sciences and establish them in their proper place on the foundation of the absolute values of the heart of God. In fact, the purpose of the first ever university in human history, the Academia, was to construct a science with God (idea of goodness) at its zenith, and this ideal will finally return to its original status through the academic model of Unification Thought.

May God's blessings and grace be upon you, your families, and your nations! Thank you.

ATHEISM AND GODISM

KEYNOTE SPEECH AT THE 21ST INTERNATIONAL SYMPOSIUM ON UNIFICATION THOUGHT

DECEMBER 6, 2009, ISSHIN EDUCATIONAL CENTER, URAYASU, CHIBA, JAPAN

I am deeply honored and highly delighted to be able to speak in the 21st International Symposium on Unification Thought. Today, the topic for the Keynote Speech is: " atheism Godism " Here, I pointed out three basic characteristics of thought to play a leading role in the world civilization.

First, the philosophy to lead the world civilization must have a firm moral root. The progress of society and culture cannot be measured in terms of technological advances or material superiority; rather, the maturity of a society and culture should be assessed by the criterion of morality. True philosophy should be able to speak for the conscience of the age with moral persuasion in times of crisis.

Second, the philosophy to lead the world civilization must be able to embrace both civilizations of East and West and create a new synthesis from them, especially bringing harmony to the materialistic orientation of the West and the spiritual emphasis of the East. Also, it should be able to harmonize the technical culture based on the analytic and logical thinking of the West and ethical culture based on the intuitive thinking of the East.

Third, the philosophy to lead the world civilization must discover God, tragically lost in the modern civilization, and adopt the absolute value centered on God as its foundation. Based on an excessive emphasis on human reason, Western civilization has dethroned God, replacing Him with a humanistic philosophy and even promulgating atheism and materialism.

I see that the philosophy that is best equipped to lead the world civilization of the future is Unification Thought, established by the Rev. and Father Sun Myung Moon. This philosophy enables us to discover the true position of God, the meaning of human existence, the social meaning of human relationship, and harmonious order of human beings and nature.

In expressing the flow of the history of human thoughts straightforwardly, a colossal struggle of the conceptions of God lies at its basis. It would not be an exaggeration to say that human history is a history of the struggle between theism and atheism, and the conflict of the conceptions of God where different views of God are vying to emerge as the sole winner.

Modern civilization can actually be said to have originated from the battlefield of fierce and bitter struggles of views of God, which vied with the God of Christian orthodoxy in one big fight to decide the winner.

Strauss, the Hegelian leftist, asserts that Darwin's evolutionary theory, and not God, is the superior theory which explains about the world more reasonably. He predicts that if Darwin's evolutionary theory were to become the basis of ethics instead of Jesus Christ, then in the place of ethics of the grace of God, compassion, love and so on, the power of the strong demonstrated in the "war on all against all" and "survival fitness" will become the new form of ethics.

In the background of scientific materialism are two premises that scientific knowledge is the most believable and verifiable knowledge and that matter is the most basic actuality in the world. Scientific materialists assert that science is the only objective and universal knowledge which is at the same time also progressive. In scientific materialism, physical phenomena, natural phenomena, and even life phenomena are explained based on mechanism, and all fields of science must ultimately be reduced to materialism.

Therefore, in scientific materialism there is no room whatsoever for God to intervene. Scientists can, without the supporting hypothesis of God, successfully establish theses in various fields of experimental science including physics, chemistry and biology, and they emphasize that, through materialism alone, with God excluded, it is possible to make successful scientific researches.

The science of today, which boasted of universal and objective knowledge under the banner of "value-neutral science," has, contrary to its original intent, revealed a value-oriented propensity. In contrast to general knowledge, in spite of the fact that science is a process of researching theories to explain objective facts, scientists carry out their scientific researches based on certain premises. This premise and prejudice is none other than materialism and atheism. However, materialism and atheism are value systems of a kind that deviates from the field of science where researches are made on objective facts. The scientists of today knowingly or unknowingly carry out scientific researches premised on value systems of materialism and atheism, and this is the real state of affairs of today's science, and the irony of it, for science emphasizes value-neutrality.

We reject the presuppositions of modern science, which lead to mechanism, reductionism and materialism, not because they threaten religion but because they are fallacious strategies which, by demolishing all metaphysics, demolishes the very science they set out to account for. Both science and religion are human enterprises, which pursue truth, arising from puzzlement about this world.

In the light of Unification Thought, we can see a prospect of harmonizing science and religion. Generally speaking one could say, 'science is value-neutral,' but this value-neutrality of science is itself a value-judgment. Maxwell equates such value-neutrality with value-blindness, or value-insensitivity.

Unification Thought is also called Godism or Head-wing Thought. But looking at it from its founder it can be called the Thought of Rev. Moon. Everything written in the Unification Thought by Dr. Lee has been reported to Rev. Moon and he has revised it and improved its content. Therefore it can be called Rev. Moon's Thought. In particular, the entire chapter of epistemology was covered by Rev. Moon. Rev. Moon is the alpha and omega of the Unification Thought and he was the one who conceptualized his own words into an ideological format. Thus the Unification Thought can be called Moonism. The Theory of the Original Image in the Unification Thought pertains to the views of God in Christianity and the Rev. Moon himself also formulated this conceptions.

Unification Thought's methodology in regard to the Theory of Original Image is, similar to scientific methodology, of a hypothetical deductive nature. Looking at it from the aspect of methodology, Unification Thought's Theory of Original Image has a hypothetical deductive system. However, same as in the case of scientific theories, for the theoretical system of the Theory of Original Image learned through intuitive knowledge to not only remain simply as a hypothetical system but to remain as the truth, it has to be verified through experiential facts.

From this viewpoint, Unification Thought's Theory of Original Image gives evidence to unmovable facts whether it is implemented in the natural universe, human history or any part of the Bible, and so it does not end as being only one of several hypotheses but instead can be said to be a theory and a truth system with a firm foundation.

In Unification Thought, the many problems that have to do with human beings, all things in the universe, society and the world, art, values, culture, history, education, science and so on, are all theories which can be deduced from the Theory of Original Image in relation to God. That is why the Theory of Original Image becomes the basis and the standard for all theories. And this Theory of Original Image is not merely a hypothetical system but it has been confirmed and verified through experiential facts in all fields of learning, and so it can be said to be an indisputable academic theoretical system and truth system.

At the 9th International Unification Thought Symposium held around this time, President of UTI Sang-Hun Lee emphasized the following point.

"As you know well, when we watch the current global situation, especially when we watch that from the viewpoint of establishing an ideal world, we can find that all spheres of culture including politics, economy, society, media, arts and so on, are collapsing as a whole. I am, at this occasion, very desirous that we provide a momentum for overcoming today's confusion by creating a revival movement centered on unification Thought. I would like to call this movement 'the Renaissance of Godism.' That is to say, I suggest that we revitalize the Unification Movement by bringing the Renaissance of Godism into existence at this occasion."

The 'Godism Renaissance' Movement presented by the President Sang-Hun Lee signifies that all scholastic world and cultures must be newly established centering on God, not materialism or humanism. God is the truth of all truths and the source of truth. The posture of learning how to love God is the beginning of all forms of learning and the fundamental of solving all social problems.

I sincerely hope that this symposium does not end in the exchange of the opinions of various scholars, but by applying Unification Thought to your own area of specialization, a turning point for forming and reconstituting new science will be achieved. It is my hope that through this symposium, Unification Thought will be developed into a movement of science. The centripetal point for shaping science is Unification Thought, and we need to tap the possibility of unifying the various thoughts with Unification Thought as the axis.

In this sense, this symposium builds up the flow of a new science with Unification Thought as the centripetal point. Hence I think this should be a gathering by which the movement of science with a new viewpoint quickens.

The papers to be presented this time are new in terms of their composition and creativity, and each bears the hard work of the presenters. I would like to request all of you to show keen interest in them and hold lively discussions on them.

WHAT IS LIFE?

CENTERING ON THE PERSPECTIVE ON LIFE BASED ON UNIFICATION THOUGHT AND SCIENCE

KEYNOTE SPEECH AT THE 22ND INTERNATIONAL SYMPOSIUM ON UNIFICATION THOUGHT

DECEMBER 4, 2010, ISSHIN EDUCATIONAL CENTER, URAYASU, CHIBA, JAPAN

I would like to express to the distinguished participants and observers my heartfelt welcome and gratitude for participating in the 22th International Symposium on Unification Thought.

My topic of keynote speech of this symposium is "What is Life?" Now is as good a time as any to redefine what life is. The traditional perspective on life and the modern scientific perspective on life cannot but clash with each other. The definitions of life presented by science, philosophy and theology are the results of their respective methodologies and imaginations. Human beings have passed the stage wherein they can cling on obstinately to just one viewpoint in defining life and dealing with matters of life, for life is something that is so complicated and mysterious that it cannot be prescribed by just one language.

Herein lies the reason why an integrated and connected research need to be carried out in dealing with life and matters of life. Modern genetics subdivide life as something that can be reduced into molecular-chemical components, whereas theology relates life to blessings from God and expand it in terms of cosmic Christology. Philosophical reason and imagination arise from this conflict between these two physical and spiritual perspectives on life, and everything comes down to the question: "Can life be divided?" This question, in turn, brings up new questions like, "Is there one life or several lives?" and "Are the genesis and growth of life gradual or rapid?" What first needs to be done in dealing with questions on life is to ostracize perspectives on life that are contrary to the essence of life, since it would only be right to make an attempt to define life after doing so.

I.   Genetic Determinism—Mechanistic Reductionism

Atomists can be said to be the first reductionists. The reason Democritus' atomism is widely acknowledged today is that it not only regarded the discontinuity of atoms from the mechanistic viewpoint, but it also presupposed that, as a general rule, every change and diversity in the natural world can be explained based on the smallest concepts like shape, size, speed and so forth. They put the case that, even when dealing with life and the mind, which have the most complicated structures ever known, in principle they could also be comprehended based on the simple concept of quantity. Galileo's dynamics was the second attempt to propose an explanation from the viewpoint of reductionistic physicalism. According to his views, the secondary qualities (color, sound, taste, etc.) could be explained by the primary qualities (which can be mathematically applied, like movement and extension).

Such concepts became the general principles supporting the unity of science asserted by the Vienna Circle. The goal of uniting science is based on the belief that all scientific truths can be defined in the precise language of sense-data, that is, the belief that they can be defined using extensive concepts under the substantial language or the natural language. If the theme of 'reduction' were to be extensively applied in different sectors of learning, an interesting question is raised. In other words, if we were to take into consideration the question of reducing linguistics using psychology, psychology using biology, and biology using physics, a very important question from the structural aspect of the different theories of modern science is implied. Logical empiricists believed that all sciences could be reduced through physics (i.e. physicalism), which is fundamentally the most universal science. In the end, however, there is still room for doubt in regard to whether the logical empiricists succeeded or failed in achieving their anticipated goal of bringing about the unity of science. Nevertheless, their ideal still remains as one of the most important themes in science and philosophy.

Bio-mechanism has been recognized from the time of Descartes as the basic perspective of modern science on life and living things. From this viewpoint arose biological determinism, which asserted that all biological phenomena including not only the functions and capacities of organisms but also the evolution of organisms followed physical natural laws. In relation to the biological determinism of today, there is a ruling paradigm in science that explains the phenomenon of life: genetic determinism. Genetic determinism regards a gene as the most fundamental essence of a living being and that the gene or the genome itself is not changed by environmental effects. To put it another way, it interprets a living thing as the inevitable result of the genome, which is the sum of genes present. This viewpoint is the perceptive basis for biotechnological researches including the human genome project. Moreover, it has been extended to social science and the general public through Richard Dawkins' concept of 'The Selfish Gene' or Edward Wilson's sociobiological ideas. Molecular biology, which tries to explain life phenomena from the molecular level of DNA, has gone beyond biology and expanded to give birth to the reductionistic world view that claims that even sociological phenomena are determined by the genes.

This world view regards living organisms as the total sum of physiochemical elements, and deems that the process of forming an organism from the nucleus through cell division, the process of forming a high molecular cell through the mutual bonding of molecules, and the process of forming a molecule through the mutual bonding of atoms are all identical. The only difference lies in the fact that, in a living thing, the combination of high molecular cells form a single organism and that it reproduces itself. The constituent in the organism responsible for this self-reproduction is the DNA. In molecular biology, the DNA is considered as the blueprint of life, and it asserts that the reproductive program of the DNA and mutations caused by the environment gave rise to the appearance of the different existing life bodies of today.

Essentially, the DNA is made up of units of nucleic acid in the shape of a 'double helix' in peculiar bonding pairs (A-T, C-G), which consist of four different bases (Adenine, Thymine, Cytosine, Guanine). A gene refers to the arrangement of sections of DNAs with special sequences and functions. The DNA code that determines genetic character is read through the messenger RNA, which is the working copy, and the mRNA is created from surrounding proteins to read that DNA code. Afterwards, proteins from different groups read the mRNA code, and from the proteins that have read that code are created single protein units called amino acids, which in turn goes through complex mutations several times and finally results in one protein. As can be seen, this multiplication process of an organism through meiosis, from proteins to cells and from cells to an organism, is only the procedure of carrying out a consistent operation according to the base sequence of the DNA compounded at the beginning. From the molecular biological viewpoint, the multiplication of all forms of living things existing in the ecosystem of today is not the purpose-oriented result that is only bound to the purposiveness. Rather, it is the result of the organic structure, which is determined based on the genetics and mutations of the original DNA, adapting itself to environmental conditions. Therefore, all living things, ranging from different types of bacteria to plants, animals and even human beings, are only the results of the realization of some of the latent possibilities in the course of the genetics and mutations of DNA.

In particular, Richard Dawkins regards all living beings including individual human beings as survival machines created through DNA or genes, and asserts that the purpose of those survival machines is to preserve their genes, which are their masters. Consequently, the actions they take to leave behind their genes for the descendant by helping those organisms that have more genes similar to theirs than others arise from the selfish gene. Genetic determinism gives rise to other questions of reductionism in molecular biology. This is because the genetic viewpoint that only tries to comprehend organisms as a form of physiochemical structure and the opinion that peculiar character of organisms are only determined by genetic actions opens the way to mechanistic and reductionistic approaches.

The reason I am criticizing genetic determinism in connection to reductionism is that it deviates from the nature of scientific theories, which should be objective and value-neutral, and is a kind of value-biased characteristic. Contrary to what is generally known, though science is a process of researching theories to explain objective truths, scientist usually begin their scientific research based on a presupposition, which in most cases are materialism and atheism. Materialism and atheism are a kind of value system. Known or unknown, the scientists of today are carrying out their scientific researches on the premise of the value systems of materialism and atheism, and this is also the irony of science which emphasizes value-neutrality. From this perspective, genetic determinism with a reductionistic character is not an exception as well.

The simple meaning of the term reductionism can be interpreted as: "A complex system is nothing but the sum of its parts and that an account of it can be reduced to accounts of individual constituents." After the existence of cells was first confirmed under the microscope, new developments in cytology were brought about, and after the efforts to understand organisms from the viewpoint of their smaller units, the DNA, bore fruit, a reductionistic approach to life became possible. The situation, however, becomes more complicated if the term reductionism is taken into consideration in relation to physicalism. Genetic determinism, which ultimately defines an organism as the cells and the DNA on the molecular level within cells, firmly adheres to the view of physicalistic reductionism, and therefore it stipulates that an organism is a substance with a physiochemical nature and in the end speaks for materialism.

In modern times, atheistic challenges from Marx, Darwin, Freud and scientific materialism are still gaining strength. In the field of biology, Darwin's theory of evolution is studied in all parts of the world in the name of scientific truth and is nurturing atheistic biology. Strauss, who is a left-wing Hegel, proclaims that Darwin's evolutionary theory is a great theory that explains the world more rationally than God. He predicted that, if Darwin's theory of evolution and not the Christ became the basis for ethics, then instead of ethics based on the God's blessings, mercy and love, the power of the strongest based on 'war of each against all' and 'survival of the fittest' would become the new ethics.

Darwin substituted God's 'creation' with 'natural selection' and asserted that organic species could evolve into other species through natural selection without the creation of God. Richard Dawkins says, "Darwin made it possible to be an intellectually fulfilled atheist" and decries creationism and 'Intelligent Design theory'. He stated that if there is a supernatural intellect with enough complexity to design something, it is only the final result created by human beings in the gradual process of evolution, and based on this definition God is only a delusion. In other words, he is saying that God is a harmful 'delusion' standing in the way of scientific imagination. Then is God really a delusion like Dawkins asserts? And would it be possible to find God in atheism, a doctrine of the theory of evolution, when God is excluded from it. I am of the opinion that, rather than making an emotional appeal by refuting the absurd claim that monkeys evolved into human beings, we should critically examine the problems of evolutionary biology, which reduced the motions and mysteriousness of life through materialism and mechanism, and make an attempt to come up with the right definition in regard to life from the perspective of the Unification Thought.

II.   Living Model - Wholism

From the viewpoint that science is the ensemble of a variety of theories, it cannot be denied that the matter of the unity of sciences, which has been traditionally presented to us from long ago, is dominated by upward causation, that is, the question of 'reducing' it using physicalistic language. What this means is that the higher and more inclusive scientific theories on the ontological level should be reducible using the lower and more basic scientific theories on the ontological level. This standpoint is premised on the belief that language and scientific theories about physicalistic sense-data, which is on the lowest ontological level, is none other than the final cause of all learning.

Nevertheless, downward causation, which is in direct contrast with reductionism, also has an important significance in modern science. Reductionism or the upward causation model and the downward causation model have adopted different world views and explanation modes. Whereas upward causation is a mode of explanation modeled on mechanism, downward causation is one modeled on the organism, that is, wholism, which is going to be dealt with in this chapter. These two viewpoints present two very different worlds to us. If reductionism or upward causation is persuasive in explaining such concepts like the mechanistic world view, materialism, and behaviorism of atomists, downward causation is more persuasive in explaining the teleological world view of organicism, mentalism, spiritualism and the like. For instance, it can be seen that, when carrying out a research on controlling the molecular activities of a cell in the biological field, the scientist tacitly adopts the downward causation model. This is because such researches in biology take as their basis the information of a higher cellular level and adopt it as the cause, and the process of controlling cells on the lower molecular level becomes their main subject. N. Chomsky, for example, devised a scientific linguistic theory modeled on downward causation by adopting the spiritualistic hypothesis of innate linguistic ability (universal grammar).

From this perspective, Unification Thought can also be regarded as a downward causation model centered on God and spiritual values. Unification Thought presumes that the mind and body of a human being is one united entity, and comprehend the former and the latter as being in the relationship of the subject and the object, the cause and the effect, and the vertical and the horizontal. Thereupon, since the mind and the body are the causational cause and effect, the causational effect is instituted by the mind, which is the subject. Therefore, the relationship of subject and object becomes the downward causation model with the subject as the cause.

To give an example, in the human body, the life of subconsciousness (mind) is the subject and the tissues and cells (body) is the object. From the causational perspective, the life (mind), the subject becomes the cause and regulates and controls the body, the object. This model of relationship between subject and object is not limited to mind and body; rather, it is the most fundamental principle that governs over all living things in the world including the relationship between God and human beings, human beings and human beings, and human beings and all creative beings. To put it another way, the existing world forms one orderly organism connected through a causational chain of subject and object. Not only the natural world but the human world also forms one orderly unified whole. As can be seen, the principle that maintains the order system of subject and object in the natural world is the law of nature, and the principle that maintains the order system of subject and object in the human world is the law of values, that is, moral law. And studying the laws governing these two worlds is a task for all fields of learning including science, philosophy and arts.

In this model, the subject takes the lead before the object logically and axiologically, and this is only because the subject has subjective, active, central and vertical values whereas the object has non-subjective, passive, peripheral and horizontal values and thus the former and the latter are able to maintain a mutual relationship. As the body cannot exist without the mind, realistic science cannot be established without values because the inevitable relationship of subject and object cannot but also be formed between values and facts.

Now, then, the question becomes clearer. Why do we criticize materialism and mechanistic reductionism? It is because the world views presented by these theories are a faulty, which cannot satisfactorily explain the existing world. In other words, materialism and mechanistic reductionism does not show the world as it is, but rather shows a reversed and upside-down version of it. And these theories present the actual circumstances of causation in a reversed and distorted form. If the fundamental cause of motion is presumed to be in matter, and the foundation for the lowest part in the causation chain is based on materialism, a clear explanation of kinetic phenomena cannot be given. According to Aristotle, the cause of motion is not the substance but the idea, that is, he asserted that it resulted from entelekeia, and when it comes to motion this mode of explanation is more rational than the materialistic one. Materialism distorts the principle of life, and explains the cause of motion in reverse. A mode of explanation for the principle of life need to be found in the wholistic or organic world view, for it is only right to do so.

In any case, the meaning of life should be found in the oneness or the unity of all organisms. In a practical sense, the dying signifies leaving life behind and being broken down into smaller parts, and the living signifies forming a wholistic organic unity in a connected network. Aristotle termed the soul or the faculty of life as entelekeia. To him, having a soul meant being alive, and he called the soul's mode of existence as entelekeia, and he said that entelekeia required the voluntary activity of the soul. Therefore, entelekeia is a faculty belonging to the soul and not matter. From the perspective of Unification Thought, love is a faculty belonging to internal character and not external form, and similarly life too should be considered as a faculty under internal character. Moreover, from the same perspective, the principle of life in motion and growth is causation from the internal character and not the external form.

Mechanism asserts that the whole is the sum of its components, but wholism states that the whole is more than the sum of its components. Because in mechanism the whole is the sum of its parts, it can be premised that the parts are the basic units of the whole. In contrast, in wholism the whole is premised before the parts. In mechanism, organisms with life are disintegrated into their physiochemical elements at the time of their death, and thus their life also comes to an end. In wholism, on the other hand, even when organisms with life perish and are disintegrated into their physiochemical elements, their life is not disintegrated, for life cannot be reduced into its material structure. As can be seen, the life of an organism cannot be reduced into physiochemical elements, and it is the subject that leads all organisms as their superior and the continuity connected to cosmic life. Therefore, the meaning of death is that the organism itself can ultimately be broken down and disintegrated into its constituents, whereas life signifies the fact that it is a continuity that cannot be broken down and disintegrated into its constituents.

Aristotle said that life is linked to eidos, and that the eidos in the individual entities are linked to the greatest eidos (eidos of eidos), that is, God

Himself, and this mode of explanation is similar to the standpoint of Unification Thought. Like Aristotle, Unification Thought also regards life as the internal character (subconsciousness) and the life within that entity as the continuity of the network connected to cosmic life (cosmic consciousness). The only difference is that Aristotle understood life metaphysically whereas Unification Though understands life in a more biological sense and thus gives it a modern meaning. To sum it up, though life is the wholistic subject governing an organism that is destined to be disintegrated in death, it is a mysterious something that does not perish even when the organism perishes. Then the question arises: if the life of plants and animals is something that does not perish after the death of their physical bodies, but instead returns to the order of comic life, is the life of human beings different from that?

Next, I will examine the wholistic world view of modern science from the perspective of Unification Thought. In classical physics, the world was regarded as a regular and deterministic model of clock mechanism governed by Newton's laws of dynamics. This view was dramatically changed with the discovery of the chaotic system. This chaotic system is extremely sensitive to the smallest external disturbance and therefore cannot be separated from the rest of the universe, and motions within it cannot be predicted accurately. In addition, Prigogine revealed that a new form of order existed in the background of this chaos, which is a naturally occurring order formed by the regular order in a delicate equilibrium.

As is seen, the wholistic world view understands the world as a unified whole that is not divided in any way. There is a limit in explaining things like forces in the molecular level, entelechy of organisms, souls of animals, and sprits of human beings as peculiar faculties or phenomena in the material world. Such spiritual phenomena have their own independent and characteristic nature, and as Plato pointed out, they come together in a mysterious bond with the material world and form one cosmic organism. This nature of spiritual phenomena is also in line with the assertions of quantum theory. From Descartes' mechanistic point of view, spirit and matter, the subject and the object are thought of as completely separate entities, whereas in quantum theory, between the two exists not an antinomy but a surprising paradox, that is, the wholistic complementarity of the observer (subject) and the observed (object).

Cosmology of Unification Thought also presents a model of biological organisms as illustrated by modern science. A perfected world is similar to a human body, in which the different cells and the parts and the whole form organic relationships with one another and thus formed a unified whole. In light of this view, Unification Thought states that the whole with parts excluded cannot exist, and individuals (parts) that are not premised on the purpose of the whole cannot exist, either. In this regard, cosmology of Unification Thought is in accord with the wholistic world view of modern science. Then let us begin in earnest to discuss the question at hand: what is life?

III.   What is Life?—What are Reducible and Irreducible?

The latter half of the 20th century can be summarized as the continuous decline of philosophy and the rapid progress of biology. Integrating physics and chemistry in itself, and unfolding its progress in zoology and anthropology, the biological theory of evolution became embedded as the integrated principle for all existing things. Furthermore, the biological theory of evolution even exerts an influence over the mentality of human beings and presents the fundamental principle for ethics and sociology. Today's DNA gene code in biology can be said to be equivalent to Plato's idea of 'the whole exceeding the sum of its constituents.'

Then what is life? There are several definitions about life. The physiological definition is that life is the faculty of a living being that makes it eat, excrete, respond to stimulation and carry out physiological functions. On the other hand, the metabolic definition stipulates life as the faculty of a living being that makes it carry out a ceaseless exchange of matter with the outside and have metabolism. Next, genetics, which is becoming more and more important with the passage of time, defines life as the reproductive function of a living being duplicating itself into another identical entity. And lastly, there is the thermodynamic definition which can overcome the drawbacks of biochemical definitions about life. It defines life as a low entropy in the existing world that is open to the free ingress and egress of energy, i.e. a special quality that continuously maintains a high order. This paper will critically examine the third and the fourth definitions that are in contrast with each other, in relation to Unification Thought, and thus make an attempt to give a new definition about life.

First, life is consciousness. As has been mentioned above, Unification Thought claims that life is individualized cosmic consciousness projected into cells or tissues. The cosmic consciousness permeates into the individual cells on the cellular level and become individualized, and this is termed as the protoconsciousness and it is the life of the cell. Just as an electric wave enters a radio and produces sound, the cosmic consciousness enters a cell or a tissue and perceives all information in it and controls its subsistence. In the end, life refers to the protoconsciousness, which is the subconsciousness with sensibility, perceptiveness and purposiveness. When God created the universe through logos, He inscribed special information in the cells of every body of living organism in the form of a material code. When the cosmic consciousness is instilled into a cell, the instilled cosmic consciousness apprehends the genetic information of each cell's DNA. Moreover, the information collected from all cells and even the smallest tissues in a human body is transmitted to the central part through the peripheral nerve. And the maintaining of order or information in the body is transmitted from the central part to the cells through the peripheral nerve. In the same way as the exchange of information in our bodies, the spiritual consciousness (mind) and the brain cells interact with each other with information as their intermediary.

Generally, life refers to the life of an organism, but in the biological sense life exists in every unit from the organism down to its organs, tissues, cells and genes. A cell is the most basic unit of a living matter that can continue to carry out all activities to maintain life, and in other words it is the small room where life dwells. By stipulating that life is the individualized cosmic consciousness in the cellular level, Unification Thought regards the life within individual cells life in an organism, and the cosmic consciousness filling up the universe as being a continuation of consciousness., Therefore, from the essential meaning of the word, life is a continuity that cannot be divided or severed, a great unity, and a pure maintenance of the unceasing exchange and transmission of information. If life is split up and divided into parts from the flow of continuity and consciousness, then it is severed and becomes dead. In this regard, life is something that cannot become divided or dead. Here we come across the einai of Neo-Platonism and the pure maintenance of Bergson. In Neo-Platonism, the einai is life with a godly essence, and it passes through a continuous process of an outflow in which it is divided into the souls of human beings and living organisms and then is returned.

To Bergson, life is a pure maintenance that is élan vital. Bergson regarded the theory of evolution, which advocated 'natural selection' based on 'coincidence,' as materialism, and emphasized that evolution is in fact the creative evolution based on the vital force of life itself. He interpreted both mechanism and teleology as determinism, which tries to draw a conclusion from the temporal relationship or its reverse between cause and effect, and criticized that they were wrongful metaphysical theories that tried to divide and compartmentalize the duration of time. He said that plants have the ability to create the energy they need to continue to exist from air, water and earth, i.e. the ability to create organic matter by absorbing carbon and nitrogen. In addition, he deemed that plants have an intrinsic inclination to evolve within itself, and thought that the vital force of life that works as the fundamental cause of mutation exists in the idioplasm of individual beings. He also believed that the vital force of life is not created by the conscious effort of individual beings but rather by a form of unconscious intuition within life itself. He says that the intellect of human beings does not understand the pure maintenance or vital force of life, and that it is only the intuition in consciousness that can perceive life and pure maintenance.

Similar to Bergson, Unification Thought also regards life as a continuation and a flow, and asserts that we can overcome mechanism and materialism temporizing the material civilization only by perceiving the living life.

As has been examined above, if life is regarded as the protoconsciousness or the subconsciousness in Unification Thought, then life cannot be reduced into material elements, and what is more, it cannot be replaced by any physiochemical elements. Though a cell is in fact made up of physiochemical elements, which is matter, the elements constituting a cell is more than matter, for there is +α in addition to matter. And this +α is something that cannot be reduced into matter, since it is life itself.

Next, let us look into scientific opinions that define life in relation to the laws of thermodynamics. In an isolated system, thermo-energy heads towards the greatest state of disorder, i.e. a state of equilibrium of unusable energy. In other words, it reaches a state of thermal death, a term presented by Ludwig Boltzmann, famous for the Second Law of Thermodynamics (law of increasing entropy). If the universe were a great isolated system, it would reach a state of annihilation, which is a state of complete disorder, based on the law of increasing entropy. Life phenomena, however, shows a different aspect from that of other general matter in the natural world. As is shown by the Second Law of Thermodynamics, it can be proved that natural phenomena advances to a state of disorderly chaos with the passage of time. In contrast, organisms always establish an order as time passes and maintain their own functions and forms within that order. In 1886, Boltzmann stated that organisms endeavored to survive to acquire entropy rather than chemical elements.

Deeply impressed by Boltzmann's insight into life phenomena, Schrodinger, who is the founder of quantum mechanics, wrote a small book entitled "What is Life?" based on his lecture given at the Dublin Institute for Advanced Studies in 1944. This book became the first step in molecular biology, and in the book Schrodinger presented questions about life from the pure viewpoint of a physicist. Schrodinger believed that living beings stored and transmitted biological information within the molecular structure. To understand about living beings, he tried to decipher the code of these molecules and thus to go beyond the level of the laws of physics of the time, that is, the laws of thermodynamics. His questions examined problems in relation to genetics on the molecular level; for instance, he questioned how the order can be created from a certain order. In particular, he wished to know how the order could be maintained within a cell, which was an assemblage of a large number of molecules and therefore statistically had a high possibility of becoming disorderly. In his book, Schrodinger presents the concept of single organisms as individual living things in a surprisingly clear light. The fact that organisms endeavor to maintain a state of low entropy signifies that there is a reversible process of building an order through the decrease of entropy in life phenomena. This reversible process of entropy is known as syntropy or negentropy. The introduction of negentropy indicates the advanced and complicated ability of organisms, which creates a complex structure from simple elements, an integrated pattern from incorporeal things, and the order from chaos. In this way, the law of negentropy is applied to living organisms, which is in direct contrast with other things, and it can be said to be matter plus another mysterious something, which has been called from old as the soul, vitality, ghost or spirit of a living being. Today's molecular biology and genetics has developed in a direction different from the one intended by Schrodinger, but the questions he posed were about vitalism, the soul and the principle of life.

Second, life is the life of individual living beings. And the universe is a vast network of the life of individual living beings. As has been stated above, life refers to the protoconsciousness in cellular units, but in the usual sense an independent living beings refers to a unified body that has been made one by the coming together of hundreds of billions of cells. From the viewpoint of Unification Thought, this unified body can be said to be a 'connected body,' in the sense that it has been formed by the coming together of many cells. Since it is an individual being distinctive from other beings, however, it can also be said to be an 'individual truth body.' Essentially, life can signify the whole life of the individual truth body, which is an independent living being, rather than the unit life on the cellular level. This is because, though the cells would still be alive, an animal cut in half cannot be said to be alive. Similarly, life has meaning only when it is in a unified body of external form and internal character, joining together the body and the soul. As is seen, a living thing signifies an independent organism that has attained the singleness or unity of life.

The most fundamental principle of life is that it is an independent living being and cannot be split up and divided into parts. For example, if a rabbit is cut up into two in such a way that its cells are not harmed in any way, the cells may be alive but the rabbit would be dead. If the life of the rabbit has perished but its cells are still alive, and it can be cloned from those cells, then we cannot but ask the question, 'What, indeed, is life?' In regard to this question, in the 1950s Nicholas Rashevsky of Ukraine introduced one of the most important formulae for understanding life. He said that applying physiological principles using mathematical models to life phenomena was not very appropriate, and asserted that a principle that shows the biological unity of organisms and the organic world as a whole should be applied instead. This is because the true form of life appears only when we have a firm grasp of a whole organism, whose cells form a tight connection network between them and perform life phenomena within themselves.

In this way, the principle of life in an organism is that 'the whole precedes its parts.' If this proposition is applied to the unit life in the cellular level and the whole life in the organic level, it can be reduced into another proposition: 'Whole life precedes unit life.' What this means is that, in regard to life in an organism, only when the whole life bringing unity to a whole organism is set down as the premise can the unit life on the cellular level have meaning. This is because whole life and unit life cannot be considered separately in an organism. This fact also coincides with our simple and sensible perspective on life, since the death of life signifies the death of whole life and not unit life. In Unification Thought, life is defined as the cosmic consciousness individualized in a given cell as the protoconsciousness, but this definition only pertains to the unit life on the cellular level and only refers to a single unit of life. In actuality, however, the unit life on the cellular level can only exist when the whole life of an individual living organism is premised. Therefore, this fact can be said to be already premised in Unification Thought.

Aristotle instituted biology as the model for his metaphysics, and he developed his speculation centering on life phenomena. Entelekeia, which is the principle of life, is already providing a cause and a purpose for life activities being carried out within the eidos of an individual organism. Human beings, who were originally animalistic beings, were bestowed reason from the God's spirit and finally became beings of reason. Therefore, human beings are beings whose purpose of existence is to nurture their reason to become more perfected and take part in the life of God, who is the pure spirit. As can be observed, for Aristotle, life is the principle of activity connecting individual human beings to the spirit of God, and it signifies realizing the purposiveness of human life.

At this point, however, we cannot but ask another question: what is this 'cosmic consciousness' filling up the universe and the 'subconsciousness' instilled in an organism mentioned in Unification Thought? On the one hand, Unification Thought speaks of life as 'autonomy and subjectivity of principle.' This definition is based on the creation of the world by God's logos, and the cosmic consciousness filling up the universe is put into organisms where it perceives all information (logos) about them, and this perception of organisms is termed as the subconsciousness. The subconsciousness on the level of individual organisms is called the '(whole) life' of an organism, and this is carried out in the identical way as the individualization of the cosmic consciousness in the cellular level into the protoconsciousness. On the other hand, logos can be said to be the 'sea of information' filling up the universe, and genes or organisms themselves are the summations of logos or 'information.' From this standpoint, life is the subconsciousness in an organism that has a firm grasp of its information. Herein lies the reason why life should not be understood as some material energy or matter.

Furthermore, in Unification Thought it is said, "The growth of living beings is based on the autonomy and dominion of the Principle, and it is the motion of life itself performed by the unity of consciousness and energy (conscious energy) latent in living things and the motion of conscious energy is the motion of life." The conscious energy mentioned above signifies a field that makes the flow of information possible, and does not refer to material energy. As is seen, Unification Thought interprets this consciousness as the principle of life, and shows that the essential life does not cling to matter but rather to this consciousness. In this way, life is manifested not in the physical body but rather in the spirit of a human being, and so has the character of an individual organism. Though it is very simple, the true meaning of life only appears when it is modeled on an organism that is an independent and unified being of internal character and external form, and consequently its ethical character could also be examined more closely.

The basic unit of life is an organism in the form of an individual truth body. Further, the model of the entire universe consisting of the interaction between individual truth bodies and connected bodies as presented by Unification Thought corresponds to the model of the entire universe. Similar to stoicism, which interpreted a human being as a microcosm, Unification Thought also asserts that the entire universe can be seen in an individual truth body who is an individual human being, but that individual truth body needs to exist and be understood in connection to other individual truth bodies. This gives an idea of what the entire universe constructed based on the interaction of individual truth bodies and connected bodies look like. Connected bodies formed from the complex structure of individual truth bodies, and the entire connected body made up of these connected bodies in an even greater complex structure is the universe, and this universe too is an individual truth body. In the universe, the cosmic consciousness (logos) filling up the universe is infiltrated in every form of existence and individualized, and these individualized consciousness, that is, the monad of life, form one individual truth body. In short, the universe is a majestic and purposeful connected structure of individual truth bodies made from the monad of life.

At this point, we are reminded of the 'omega point' of Teilhard de Chardin, who was a Jesuit and a paleontologist. It is interesting to note that 'omega point' is similar to the concept of Unification Thought that modeled the universe as 'a human being who has perfected the purpose of creation.' He combines the scientific views of the theory of evolution and optimism of the Christian faith, and thus brings into accord the perfected universe, the final destination of the evolutionary process, and the cosmic Christ. Furthermore, he describes the birth of life, a unique phenomenon on this planet, and the humanization process of consciousness and conception, which is the destination of life. He also asserted that the final destination of evolution is the sanctifying process of the cosmic Christ mentioned earlier, and that their point of union is the 'omega point.' This attempt lies up for itself new difficulties in giving an explanation by unreasonably relating geological evolution with teleology in Christianity, but because it dealt with the birth of life and the humanized model of the universe, it deserves to come under close scrutiny from the viewpoint of Unification Thought.

Last, life is the self. At this stage, we cannot but hesitate a little, and pay attention to the result of the discussion worrying that there might be a jump of logic. Nevertheless, it will facilitate a more reasonable explanation than the logical burden of theological interpretation, which stipulates life as the essence of God.

Unification Thought defines life in the cellular level (protoconsciousness) as the 'lower-dimensional mind' and understand it as the subconsciousness with sensibility, perceptiveness and purposiveness ("New Essentials of Unification Thought," 2006, 407). And it also believes that when God created the universe through the logos, "He inscribed all the information (i.e. logos) pertaining to each living being in the cells of that being in the material form of a code for the maintenances of its species from generation to generation." That code is the genetic information in the DNA, and the genetic character of genes become different according to the arrangement of the four kinds of bases mentioned earlier, namely Adenine, Thymine, Cytosine, and Guanine. Life (subconscious) has the ability to perceive that genetic information. Proteins can read the genetic information in mRNA, which is the working copy of DNA, because life has the ability to perceive through its cognitive activities. DNA is only the carrier of genetic information, and proteins that can read the information in DNAs are also carriers of life.

Therefore, life infiltrated in all the cells including DNAs and proteins has the preceding ability to read not only the genetic information in DNAs but also all other kinds of information about the cellular organization, structure, and so on of a given organism. Hence, life precedes the DNA. This concept can be understood in the same vein as the philosophical proposition that 'consciousness precedes existence.' In other words, life (subconscious) is the preceding consciousness that can perceive all information in all organisms, and so it precedes the DNA. Here, we can also reaffirm the principle of life that says 'the whole precedes its constituents.' The organic model of Unification Thought has instituted that, in the relationship between the whole purpose and the individual purpose, the whole purpose precedes the individual purpose. If this concept is examined more closely in the embryological light in relation to the theory of evolution, another interesting, age-old question comes up: 'Which came first, the egg or the chicken?' This question is translated in relation to the evolution of organisms into the question: 'Which came first, the DNA or the living body?' And in terms of molecular biology, it becomes the question related to origin of life: 'Which came first, the DNA or the proteins that read the genetic information in DNA?' Genetic determinists or evolutionists will support the former question in regard to this issue, whereas vitalists or 'intelligent design' theorists will support the latter.

Discussions on these questions are out of the sphere of this paper, and so we will not deal with them here. Nevertheless, the plain fact of the question of living beings is that life precedes the DNA, and therefore, whether it is in the logical sense or the embryological sense, life does precede the DNA. As can be seen, life functions before the constituents and has the ability to control changes in the parts as a whole. And since it goes beyond the systematic structures of the individual living being, life can be said to be transcendental. Life is not the summation of the parts but instead it carries out integrated activities in the organism as a whole. This ability of life to control the whole is stored as the material seed in the form of DNA. For this reason, DNAs cannot be regarded as life in themselves; rather, they are the vehicles of life in a material form. This is why no matter how many times a scientist clones DNAs he is only creating the external form of an organism. What a scientist can create is not life itself but the carrier of life.

From the standpoint of Unification Thought, life is more than the sum of DNAs or the constituents of a living organism, for life is the consciousness that controls the organism as a whole. In short, it is the subconsciousness. Life is not included in the domain of a living organism, and in this regard it is transcendental. If, as has been said above, the autonomy of life operates transcendentally in an organism, then what is its relationship to the soul? Does a human being really have an independent soul that can be called his transcendental self? Let us now begin in earnest to examine this question in relation to the transcendence of life.

The viewpoint of Unification Thought is not based on spiritual or materialistic monism, any more than it is based on spiritual or materialistic dualism. According to Unification Thought's structure of Original Image, all beings exist having the dual characteristics of internal nature and external form. The internal nature and external form are in a relationship of subject and object, through which one governs over the other and the other is governed, and one plays a leading and active role while the other plays a passive role. When the relative relationship of internal nature and external form is examined, centered on life, human beings can be distinguished into two unified bodies: the unified body of the consciousness (mind) and the organism (body) and the unified body of the spiritual self and the physical self. In the case of the unified body of the consciousness (mind) and the organism (body), human beings play a subjective role in which the consciousness (life) governs the organism with autonomy. This case is similar to other organisms, and has already been explained above. In the second case, however, if we were to define the dual characteristics of a human being as the unified body of the spiritual self and the physical self, then a new question arises: what is the relationship between life and the spiritual self?

What, then, in the perspective of Unification Thought, is the higher-dimensional mind on the organic level, which is the counterpart of the protoconsciousness, the lower-dimensional mind on the cellular level? When viewed on the cellular level, the life of the cell or the protoconsciousness perceives the structure and details of the cell and is connected to the life of the organism on the physical-self level. Thereupon, life on the cellular level can be termed as the lower-dimensional mind or the psyche, and the life on the physical-self level of the organism can be termed as the higher-dimensional mind or the spiritual soul. The higher-dimensional mind on the organic level is connected to the cognitive activities of a human being. In relation to the cognition of a human being, the higher-dimensional organic life (subconscious), that is, the cognitive ability of the spiritual soul functions in a similar way as the perceptive ability of life on the cellular level.

Whether it is the mind (consciousness) instilled in the minor cells or the mind (consciousness) instilled in the brain cells, the original form already exists inherently in all the different kinds of the mind (consciousness). This original form is the inherent and precedent standard in the mind by which all external empirical objects are judged. As can be seen, the precedent original form and the responding empirical content are the cognitive activities of a human being. In a way, as Plato asserted, the cognitive activities of human beings could be the actions by which human beings are reminded of the original form inherent in the flow of consciousness, which is life itself, through external experiences. Moreover, in another way, as is advocated in phenomenology, the mind could be the noesis of consciousness that has the precedent noema within itself and asks for the meaning of what it perceives. In any case, Unification Thought talks about the subconsciousness of the mind that leads the cognitive activities of human beings on the organic level, and this consciousness is defined as life. As is seen, human beings are conscious beings who carry out cognitive activities with consciousness, as well as the self-transcendental subject who manage and control their physical selves through their consciousness (life). Generally, this self-transcendental subject is called the spiritual soul or the spiritual self, and it is the self-transcendental monad that rises above the physical self. We call this the 'self.'

As has been examined above, the spiritual self in Unification Thought can be called the 'self' in the common sense of the word in relation to the essence of human beings. It is the self-transcendental consciousness, the autonomy and life itself, which governs over the survival and reproduction of all living organisms. Modern brain physiology recognizes the term consciousness as the integrated and dynamic characteristic of the process of brain activity and the central constituent of brain activity. R. W. Sperry deemed that the subjective experiences in the consciousness is the determinant factor in the functions of the brain, and thought that they had a dominant influence in controlling the process of physiochemical occurrences taking place in brain activity. In a sense, the mind in the brain moves matter, and this is similar to the organism controlling its constituents like the organs or the cells, or the molecule controlling the molecular course taking place in its electrons. It is not different from the practical perception that objective science and its theories employ to explain the brain, that the conscious mind controls actions. The 'conscious mind' stated above is in itself life with autonomy, and refers to the 'self,' that is, the self-transcendental monad, which is the transcendental subject controlling brain activity and the counterpart to the physical self. Brentano explained that the essence of self-consciousness is the ability to distinguish between the perceiving subject and the perceived object, and mentioned intentionality as the principle of that self-consciousness. In the field of physics, Heisenberg also emphasized that there should be a branch of physics that deals with both the observer (consciousness) and the observed. Gilbert Ryle defined the mind as the process of bending the spirit. Eccles, a brain physiologist, clearly distinguished between the brain and the self-conscious mind and thus made an assertion of the interactionism of mind and body.

In the book he wrote with Popper, 'Self and Its Brain', Eccles showed through experiments that, whenever the consciousness was in action, it 'paid attention', and this phenomenon of concentration was connected to the nerve phenomenon of the brain. He opposed the identity theory, which identified conscious process with nerve activity, and regarded the self that causes cognitive phenomenon and the brain as independent entities. In explaining about conscious concentration human beings employed when they tried to think up a word just out of grasp (when speaking), he showed the time gap that occurred between the cognitive phenomenon of the self and the physical nerve phenomenon through experiments. This is significant in which it revealed scientifically that the conscious subject or the self and the brain existed independently from each other and that the self and the brain interact with one another.

If the self we have dealt with above is the end of life and the whole of the life network, then a personalistic model of life and personalistic ethics could be suggested. However, the life of the self forms a greater network in relation to the whole purpose, and so ultimately a model of life and a model of life ethics need to be presented in continuity with cosmic life. That model should be a wholistic model of organisms related to the whole network, and furthermore, a model of ethics befitting its purposefulness.

IV.   Towards a New Model of Life Ethics

We have examined the perspective on life centered on Unification Thought and science. In relation to this symposium, the contents discussed above are preliminary examinations of life ethics, and I believe they will play a fundamental role in establishing a new life ethics. This is because, just as anthropology and ethics can be considered to be a pair, life and life ethics are also concepts that cannot be considered separately. As space is limited, I would like to conclude this paper by making a suggestion about the construction of a desirable model of life ethics in connection to life from the perspective of Unification Thought.

First, in considering life ethics in relation to life, we should not ignore the results of researches carried out in modern biology. Though some scientists interpret life from the physiological viewpoint and thereupon prefer materialistic hedonism and utilitarian ethical views, such physicalistic dogmatism and materialistic ethics cannot be instituted as life ethics. Nevertheless, it is not possible to ignore objective biological facts in laying down the basics of life ethics, because, though life ethics is not deduced from biology, life ethics should be instituted in relation to biology. For instance, such things like DNA recombination technology, test-tube baby operations related to abortion, or the standard of biologically determining brain death, pertain to both. Though these results of biological researches are not requisite in laying down the foundation stones of life ethics, they are still the questions that need to be answered first.

Second, in creating a new life ethics, we need to embrace and rise above the understanding of traditional views on humankind and the personalistic ethics based on them. When dealing with life ethics based on Unification Thought, we cannot cling to the traditional ethical model and insist on following it. For example, utilitarian ethics clashes with teleological ethics, and Kant's moral laws clash with Aristotle's ethical theory of virtues. And in connection to environmental issues, human-centered ethics clash with ecological ethics. In dealing with life ethics, we cannot place emphasis on just one side of traditional ethics and adhere to it. However, we should not overlook the traditional ethical model in establishing life ethics in Unification Thought. The traditional personalistic ethics on life enlightens us about the dignity and value of human life. Health and life, ethics of virtue, ethics of duty, and human responsibility are some of the keywords of the traditional personalistic ethics. This personalistic model about life is related to the interpretation of life as the 'self' mentioned above, and can be said to be an ethical model deduced from that interpretation.

Third, a new life ethics involves the process of presenting an ethical model based on Godism. Being premised on God! This may be possible when in dealing with religion, but there will be difficulties in discussing the matter in terms of science and reason. It might be possible to do so from the viewpoint of natural theology, but at any rate it will be a difficult subject to discuss rationally. Fortunately, there is a way to approach the issue of applying life ethics without premising God. This is the way of Shimjung ethics. Shimjung ethics is an appropriate model in establishing life ethics, and the two are in conformity. In Unification Thought, when life is defined as consciousness, the cosmic consciousness that is set as the premise is the sea of life, the sea of information, and the sea filled up with love codes. From this standpoint, Shimjung ethics is more than eligible to become the model of life ethics. However, there is the fact that the principles of Shimjung and life are altruistic whereas Dawkins' genes are 'selfish', and this is where the former clashes with modern biology. Is life ethics selfish or altruistic? These questions are the issues about life ethics that are related to Unification Thought and that still remain for us to solve.

THE UNITY OF SCIENCES AND UNIFICATION THOUGHT:

TOWARDS EXPLORING UNIFICATION THOUGHT ACADEMIC DISCIPLINES

KEYNOTE SPEECH AT THE 23RD INTERNATIONAL SYMPOSIUM ON UNIFICATION THOUGHT

DECEMBER 10, 2011, ISSHIN EDUCATIONAL CENTER, URAYASU, CHIBA, JAPAN

I would like to start by giving my most heart-felt welcome to honorable guests, and participants.

Unification Thought is intended to systematize the thought of Rev. Sun Myung Moon and present it in an appropriate order. Unification Thought aims to realize a world in which all of humankind can serve God as one great family and bring about a peaceful world. In this respect, it is meant to serve to bring about world of peace, complete liberation, and true love through a process of reconciliation and unification. This intent may provide insight into why Unification Thought is also referred to as Godism or Headwing Thought. Godism refers to the fact that this thought has God's truth and his love at its core, and Headwing Thought refers to the fact that it is neither right-wing nor left-wing in its approach, but instead embraces the two by considering them from a higher perspective.

Unification Thought, however, does not only intend to reconcile and unify democracy and communism, which have been locked in ideological confrontation, but is presented as a system that can offer solutions to confusion affecting the human and social sciences, the natural sciences, and the arts. The unity and harmony it brings to the various theories competing in these fields must be based on the fundamental principles of the universe.

Rev. Moon has stated that world leaders of the Unification movement must be armed with at least three basic theories. They are the Divine principle, Unification Thought, and VOC theory. The Divine principle takes on the structure of Christianity's systematic theology and the Unification Thought and VOC theories take on a more philosophical and ideological contour but they all come from Rev. Moon's words. The Divine principle is what we received when we organized Rev. Moon's words in the form of systematic theology. When we transformed his words into a more ideological context we got the Unification Thought VOC theory. Therefore we can see that all these theories have come from the words of the Rev. Dr. Sun Myung Moon.

Religion all originated from one God. But in the historical course of their formation, each religion lost its original purpose and multiplied hatred and anger towards other religions due to the limitations of the age, dogmatism, and stubborn insistence on one's doctrines. Therefore it is important that man break away from dogmatic religion created by man, break away from the prison of religious ideology, and return to the spirit of its origin: true love.

Philosophy and theology have this point in common that they both regard God as the object of their researches, but they differ in that the study in philosophy is based on reason whereas the study in theology is based on faith. In this way, theology pursues revelation through faith whereas philosophy pursues the truth through reason. In this regard, since Unification Thought is not premised on faith, it cannot be said to be a revealed theology based on revelation. However, as long as the subject dealt with in theology is studied within the scope of reason, it can become the subject matter for the research of Unification Thought. Unification Thought gives attention to the universal ideology and teachings given by Rev. Dr. Sun Myung Moon in his speeches, and tries to systematize the truth disclosed through our reason.

Scholars of the world express surprise at the fact that Unification Thought deals not only with theology or philosophy but also all sciences in general including humanities, social science, natural science and artistic fields. Moreover, they are even more amazed to learn that it presents solutions in all of the fields. Apart from whether Unification Thought is right or wrong, people express incredulity at the fact that it deals extensively with all fields of academic disciplines.

With the increasing success of science and science-based technology conflicts arose between organized religion and science. With Copernicus, who dethroned the earth as the Center of the Cosmos, Galileo Galilei was persecuted by the Church. Science seemed to give no room for a creator-god to directly intervene in the functioning of the universe. Science's headstrong style is illustrated by the story of Laplace developing a theory of the cosmos, and late twentieth century science has combined Laplace's ambition and Darwin's ideas to produce a picture of an evolving universe endowed with no purpose and developing without any specific divine intervention. Man is not necessarily the crown of creation, but just one of the products of evolution.

We reject the presuppositions of modern science, which lead to mechanism, reductionism and materialism, not because they threaten religion but because they are fallacious strategies which, by demolishing all metaphysics, demolishes the very science they set out to account for. Both science and religion are human enterprises, which pursue truth, arising from puzzlement about this world. In the light of Unification Thought, we can see a prospect of harmonizing science and religion.

Scientific knowledge is essentially objective knowledge, which has been perceived to eliminate the subjective element of observation. Therefore scientific knowledge is a knowledge derived from the reasoning mind of man. Value judgment, metaphysical theory and aesthetic consideration may be merely personal, but they may also be supra-personal. Religion, metaphysics, ethics, and the arts are objective in the sense that they seek both to understand and know about the world. Scientific theories are in principle testable and therefore reasonable, while metaphysical theory can be repeatedly redefined and criticized. Spiritual knowledge is knowledge about the spirit as object which can be gained by the reasoning mind of man.

In the Unificationist view, science and religion have a common share in the pursuit of truth. The object of scientific pursuit is external truth in order to move mankind from outer ignorance to outer knowledge, while religion searches for internal truth to move from inner ignorance to inner knowledge.

Scientific knowledge is characteristic of pure knowledge orientation, whereas Unification epistemology is characteristic of knowledge orientation together with a purpose driven end (telos). Paradigmatically, though western scientific epistemologies are fact laden, the Unificationist one is value laden. It may be correct to say that Unification epistemology is characteristic of soteria oriented teleology.

It is my hope that through this symposium, Unification Thought will be developed into a movement of science. The centripetal point for shaping science is Unification Thought, and we need to tap the possibility of unifying the various thoughts with Unification Thought as the axis.

Then is Unification Thought a philosophy, a faith system, an ideology or an academic theory? The more the study and discussion on Unification Thought is carried out by world scholars, the more the question of the nature of Unification Thought comes up. Unification Thought's understanding of God does not try to prove the existence of God epistemologically or ethically, but has its aim in knowing clearly the true form of God. Rev. Dr. Sun Myung Moon, after studying in detail the all things in the universe and the central contexts of the Bible, has come to the conclusion that God must be a God of dual characteristics, and after implementing this to nature and the universe and history and the context of the Bible respectively, he has been confirmed in his belief that his conclusion was right. This method used in Unification Thought's understanding of God can be said to fall under the methodology of academic theory, similar to the hypothetical deductive method used in science.

Scientific theories commonly used in science, such as Newton's theory of dynamics and Einstein's theory of relativity, basically have a hypothetical deductive nature. The scientific theories of today are not derived from a process of induction based on experience, but instead it has come to be widely accepted that these theories are derived in a hypothetical deductive method through the scholar's intuition and imagination. Scientific theories are made based, not on "experience" and "induction," but on "reason" and "deduction."

As can be seen, scientific theories fundamentally acquire a hypothetical deductive nature. In this regard, religious truth and scientific truth shares a common character in the psychological aspect. The life of Abraham, spent in pilgrimage to some place or other in accordance with revelations or religious intuition, or the life of scientists spent in presenting some theory or other in accordance with their scientific intuition, are both based on intuition, reasoning, imagination, and belief that it will be so in the future.

Unification Thought's methodology in regard to the Theory of Original Image is, similar to scientific methodology, of a hypothetical deductive nature. The basic concepts of the Unification Thought are simple. Sungsang, Hyungsang, Yang and Yin, Give-and-Receive Action, 4-position base etc. But these are the keys that will solve those fundamental problems that any thoughts so far could not solve. Looking at it from the aspect of methodology, Unification Thought's Theory of Original Image has a hypothetical deductive system. However, same as in the case of scientific theories, for the theoretical system of the Theory of Original Image learned through intuitive knowledge to not only remain simply as a hypothetical system but to remain as the truth, it has to be verified through experiential facts.

From this viewpoint, Unification Thought's Theory of Original Image gives evidence to unmovable facts whether it is implemented in the natural universe, human history or any part of the Bible, and so it does not end as being only one of several hypotheses but instead can be said to be a theory and a truth system with a firm foundation.

In Unification Thought, the many problems that have to do with human beings, all things in the universe, society and the world, art, values, culture, history, education, science and so on, are all theories which can be deduced from the Theory of Original Image in relation to God. That is why the Theory of Original Image becomes the basis and the standard for all theories. And this Theory of Original Image is not merely a hypothetical system but it has been confirmed and verified through experiential facts in all fields of learning, and so it can be said to be an indisputable academic theoretical system and truth system.

It is my hope that through this symposium, Unification Thought will be developed into a movement of science. The centripetal point for shaping science is Unification Thought, and we need to tap the possibility of unifying the various thoughts with Unification Thought as the axis.

Thank you very much!

THE PHILOSOPHY OF UNIFICATION THOUGHT FOR VALUES EDUCATION

Our society is now troubled with the questions of delinquency, school violence and the degradation of the sexual morality of youth. Schools, having no proper theory of education, do not teach youngsters what they should follow through their lives, but have become degraded to a place where teachers and students (and parents) buy and sell knowledge, where students do not respect their teachers, and teachers have lost their sense of authority and enthusiasm for their students.

Yet, a suitable theory of education to overcome this confusion is not to be found anywhere, and contemporary education has lost its sense of direction. To bring this situation under control and come up with a clear vision for future society, the need for a new values education has become pressing. This is the very reason for the Unification Theory of Education,[1] the Values Education of Unification Thought to come into existence.

Theories of education usually have two aspects. One is philosophical and deals with the basic principles of education and the other is scientific and deals with the objective facts of education. According to Unification Thought, the former is called the sungsang aspect of education, and the latter is called the hyungsang aspect of education respectively. Here the sungsang aspect of education, that is, the philosophy of education will be summarized.

---

[1] *Essentials of Unification Thought* (Japan; Unification Thought Institute, 1992), p. 168

I.   Values Education based on Unification Principle

A) The Three Great Blessings and Three Great Ideas of Education

i) *Educational Meaning of the Three Great Blessings*

According to Unification Thought, education can be described as the process of raising children to attain resemblance to God. God created man and woman to resemble Him (Gen. 1:27). After God created man and woman, He gave them the three blessings to grow, to multiply, and to have dominion over all things (Gen. 1:28). This becomes the foundation of values education. To resemble God is to resemble the Divine Image and Divine Character.[2] In short, for man and woman to resemble God is to grow to resemble the relationship within the Divine Image, that is, the sungsang and hyungsang, Yang and Yin, and individuality of God, and also to inherit completely the Divine Character, namely, the Heart, reason-law, creativity, and so on, of God.

As mentioned above, when God created man and woman, He gave them blessings, saying, "Be fruitful, multiply, and fill the earth and subdue it." Here the first item, "be fruitful," means to attain a personality resembling the perfection of God; "multiply and fill the earth" means for man and woman to become husband and wife and multiply children so as to resemble God in His creation of children, man and woman; and finally "subdue the earth" means to resemble completely God's dominion over creation through taking care of all things. In short, to resemble God's image means to fulfill God's three great blessings of perfection (the first blessing), multiplication (the second blessing), and dominion (the third blessing).[3]

For people to resemble God's perfection in the first blessing means the fulfillment of individuality through minds and bodies united in oneness in the unification of divine mind and physical mind centering on Heart. To resemble God in His creation of children in the second blessing means that man and woman accomplish a family with the qualification of man and the qualification of woman in harmony just as the Yang and Yin of God exist in harmony. To resemble the harmony of the Yang and Yin of God is to fulfill the man's duty as a husband and the woman's duty as a wife, and this harmony then develops into and exists on the level of the tribe, community and nation. To resemble God's nature of dominion means to accomplish the nature of dominion, that is, to attain the ability, centering on true love, to have dominion over both human objects and material objects, in a manner resembling God's creativity; He created man and woman including all

---

[2] Ibid., p.169
[3] Ibid., p.169

things with His Heart to realize True Love. What is worthy of special no-
tice in the Unification Theory of Education is that three great ideas of education originate in the three great blessings of God.[4] These ideas are of
those of the completion of individuality wherein learners are educated to
become men or women of character able to manifest true love, the completion of the family wherein are educated the attitudes of husband and wife
centering on true love, and the completion of dominion wherein people are
educated in the proper dominion of creation and all things centering on true
love.

ii) *Three Great Blessings and Universal Ideas of Education*

Prior to this subject, let us briefly compare the three great ideas of the
Unification Theory of Education with traditional philosophies of education.
Many educational philosophers have set the goals of education on building
up moral humans or characteristic humans. Socrates (470-399 B.C.) developed idea of the soul as the intellectual and moral force which made a living man a real person. He considered that the care of soul was the highest
human activity and that the individual was important.[5] Plato also asserted
that purification of soul or intellect is the aim of dialectic art and learning,
so we should understand that the art of education is to be concerned with
this kind of purification of soul.[6] I. Kant (1724-1804) also set the goal of
moral education in "the effort to make humans worthy humans". And his
moral ideal, as will be seen later, was the greatest possible approximation
to complete virtue, to the holy will of God.[7] E. D. Spranger (1882-1963) in
the modern age advocated 'magic of soul', which is essential of life,[8] saying "to keep your soul pure more than anything else" is the only end of
education.

The rationalism advocated by Plato and Kant regards the highest goodness of virtue as an ideal to be grasped by reason, and believes that reason
can purify and teach the realities and nature of human beings in a worldly
life which is affected by human sentiments and impulses. It insists that the
power of reason is essential to education. Therefore, these thinkers put extreme reliance on teaching ability and advocate the external functions of
education. But there are not a few theories criticizing this view, one of
which is the concept of "nature" of J.J. Rousseau's (1712-1778). He argued
that what we need is already within us naturally, and rejected the traditional
education of corrupt society. He felt that external education only gets in the

---

[4] Ibid., p.174

[5] Meg Parker, *Socrates and Athens*, (Macmillan Education Ltd. 1973) p.26

[6] Edit. Edith Hamilton & Huntington Cairns, *Plato*, (Bollingen Series LXXI,
Princeton. 1973) p.973

[7] Frederick Copleston, *History of Philosophy - Kant*, (The Newman Press, 1961)
p318

[8] The Encyclopedia of Philosophy vol.8, (Macmillan Publishing Co.1967), p.1

way of our own inner nature.[9] Accordingly, he set the aim of education on the development of the sensitivity of faculties and reason, which we have had from birth, before they are spoiled by socially acquired habits, and he denied the benefit of external education, saying, "Not to educate is the best education."

We cannot, however, find any moral goodness for eternity or universal validity from Rousseau's education of rationality or the rationalism that relies on human reason. The character of nature in naturalism exhibits infinite variety and rationalism cannot set up any ethics that can be recognized universally. Such philosophers say at the most that it is good and moral when one is in accord with the moral standpoint of one's society and era. Thus ethics changes according to the age and social situation.

Therefore, from a logical point of view, a universal education goal valid for all ages and places, that is, a concept of universal goodness should be established centered upon God. F. Froebel (1782-1852), representative of the educational circle in the 19th century, turned his attention to this point to systematize the philosophy of education centering on the divine nature.[10]

He thought the unity of nature and human beings can be possible only through God, for there is an inseparable relationship bringing together nature, mankind, and God. Since all things come from God and are prescribed according to God's Will, nature and human beings reside and live and continue to exist in God and through God. In addition, he insisted that the aim of education is to nurture the divine nature at the core of human beings so that it manifests in all directions and can be realized by human beings in the midst of human life.

We also recognize the importance of the Unification Theory of Education from the viewpoint of making up the universal, unchangeable theory of education concretely out of the three great blessings of God. The idea of the three great blessings for humans to fulfill through certain periods of growth can be said to be the real completion of human nature and the object of education. Accordingly, the defects of naturalistic and rationalistic education can be solved and a universal theory of education established.

B) 2. The Processes of Growth and Human Responsibility

It takes certain periods of time for human beings to grow to resemble God's image. These are the three stages of growth—Formation, Growth, and Completion. Growth means a process of coming to resemble the aspects of God's personality, God's harmony of Yang and Yin, and God's

---

[9] J. J. Rousseau, tr. Barbara Foxley, *Emile* (London; L.M. Dent & Sons Ltd., 1974), p.5

[10] F. Froebel, *The Education of Man* (Clifton; Augustus M. Kelley, Publishers, 1974), p.10

creativity.[11] Though man's physical body, like all things in nature, grows through the dominion and autonomy of the Principle itself, the spiritual body grows by accomplishing its own portion of responsibility. Growth by accomplishing one's portion of responsibility means that humans develop their character through responsibility and effort. To attain this goal people try to participate in the Heart of God through freely following norms and principles. Here let us consider the educational meaning of the portion of responsibility in values education.

Education in a sense means the exertion of all physical and spiritual influences so as to have a good effect on the development of a human being. The influences we receive in our family or society unconsciously can be said to be education. Education, however, in the strict sense of the word, is attained through intentional and systematic methods. In this sense the influences we receive from nature and society, or the simple information we come into contact with in newspapers and on TV cannot be regarded as education. Education means the effort of mature persons to guide and bring up immature persons over a certain period of time. On the evidence that 教育, "education" in Chinese characters means both guidance and whipping from outside as shown in the character 教, while cultivating and developing the inner nature and abilities of the learner through individual initiative as shown in the character 育, the educational philosophy of Spranger, can also be seen as correct. For Spranger, education can be realized through the interaction of objectified cultural values and subjective individual dispositions towards personal experience.

For the Unification Theory of Education, education can be realized on the basis of human relationships through which parents teach and nurture their children in the process of growth in order reach the goal of perfection. Therefore, the effect of education is not attained from the one sided teaching of parents, but is attained when the children obey their parents, open their eyes, and put the teachings of their parents into practice. As Adam and Eve, the human ancestors, should have fulfilled their portions of responsibility by obeying God's commandment, children should obey and follow their parents as their portion of responsibility.

O.F. Bollnow (1903-), a German pedagogue, in his book Existenz Philosopie und Pädagogik, referring to the importance of education for parents to admonish children to behave well or for teachers to warn students not to keep bad attitudes in lessons, argued that an admonition is not a command to exact one-sided obedience, but a teacher's act to promote what he should teach through students' free will. In Bollnow's sense, Freedom of decision

---

[11] *Essentials of Unification Thought* (Japan; Unification Thought Institute, 1992), p.173

is possibly only on the basis which human reality is subject.[12] Spranger dealt with the student's awakening as the most important thing in education in a way that is almost the same as the portion of responsibility of in Divine Principle. Referring to education as a cultural process that not only passes on culture from one generation to another, but also goes deeply into the spirit, he explained the function of awakening influences down to the bottom of the spirit. Therefore, education in the true sense can be realized through the interrelationship between the content of education given by teachers or parents and the response of students or children.

II.   The Three Forms of Education

Education is to foster children to resemble God, and a process leading them to resemble God's Divine Image and Divine Character concretely. Accordingly, to understand God's Character -- His Heart, Logos, and creativity—we should understand sungsang and hyungsang, yang and yin, and Individual Image from the beginning. The fundamental questions of who God is and what relationship He has with us should be explained as the foundation for education of Heart. First of all, let us observe the sungsang and hyungsang of God in relation to the education of Heart.

When we want to know the mind of any person, we observe appearance, words, and behavior. And when we want to gain knowledge of a historical figure we cannot meet directly, we study the works or contributions left behind for posterity, and look into the life achievements, for such works or achievements tell us a person's character, thought and natural disposition. In the same way, we can perceive the nature and will of god through observing nature, that is, God's work.

The Bible says." Ever since the creation of the world his invisible nature, namely, his eternal power and deity, has been clearly perceived in the things that have been made" (Rom. 1:20). In this way, though we cannot see God, we can grasp the nature of God clearly through looking into the things in nature God has created.

All beings consist of the two aspects of inner character and outer form. The invisible inner nature is called the sungsang of the being, while the visible outer form is called the hyungsang of the being. As sungsang and hyungsang represent the relative aspects of the same thing, the hyungsang may be called the "second sungsang." Thus, we can call sungsang and hyungsang together "the dual sungsang."[13]

---

[12] The Encyclopedia of Philosophy vol.1 (Macmillan Publishing Co.1967), p.333

[13] The Holy Spirit Association for the Unification of World Christianity, *The Divine Principle* (Seoul; Sung-Hwa Publishing Co. Ltd, 2004), p.19

Sungsang is the internal nature and functions as the mind, while hyung-sang is the external form as body. People also consist of two parts, the invisible mind and the visible body. When we look into the world of animals, plants and minerals closely, we can see there is an aspect of sungsang like the human mind in addition to the hyungsang, the body or form. The instinct of animals, the directional character of plants, and the disposition of minerals are manifestations of their sungsang aspect. For example, the nature of migratory birds, which return home after traveling thousands of miles, and the amazing instinct of fish, and birds to preserve their species in a hard struggle for existence is marvelous. Similarly, some scientists have reported that plants grow better when they are exposed to beautiful music.

Since all things are resultant beings which have come from the first cause God, their sungsang and hyungsang cannot be explained without the recognition that they also originated in the causal being. Hence the First cause, God, too, exists as absolute subject with the dual characteristics of sungsang and hyungsang. In short, the sungsang and hyungsang of all beings are created resembling the essential sungsang and essential hyungsang of God. Therefore, God is the harmonious subject of the essential sungsang and essential hyungsang. In other words, we can say God expresses Himself as the being of harmonious sungsang and hyungsang subjectivity.

What is the relationship between God and man? People were created resembling God's image in form, while the things in the universe were created resembling God's image in symbol.[14] The Sungsang and Hyungsang of human beings encapsulate all the sungsang and hyungsang of animals, plants and minerals. God created humans as the substantial microcosm to dominate all things in the universe. In short, human perfection means resembling the image of God perfectly. When we generally call human beings the lords of all creation, or microcosms, this refers to the special status of humanity created as the manager of all things in the universe.

The Bible says God "made man in His own image"(Gen. 1:27), which means the original man was the being to manifest God's image completely. There is an ethnic religion in Korea which postulates a human-centered thought, saying, "Man is the very God," which, however, is a little different from the view of humanity in the Unification Principle. There it is taught that the relationship between God and humanity can be analogized to the resemblance between father and child.

In the Bible, when the disciple Phillip asked Jesus to show him God, Jesus answered why he asked Jesus to show him God when they had already seen him. He concluded by saying, "The Father is in me, and I am in the Father." The meaning of this verse could be translated many ways, but the plain truth is that God the Father expresses Himself through the son

---

[14] Ibid., pp.19-20

Jesus. The first revelation that Rev. Sun Myung Moon received in his bitter struggle to find the truth is that the relationship of God and people is 'the relationship of Father and Child.'[15]

The relationship between God and the universe is the relationship between subject and object, cause and result, vertical and horizontal. The relationship between sungsang and hyungsang is also the relationship between subject and object, cause and result, and vertical and horizontal. Accordingly, looking from the perspective of the nature of relationships, that between God and man is the same as that between the mind and body. As the mind and body come to resemble each other, so humans become the beings who resemble God the most. When we hurt ourselves physically, our mind is the first to care. Whenever we do good things or bad things, our mind knows it first. Our minds always try to lead us to do the right thing, like our parents and teachers. We can say that God's love for people is the same as the mind taking good care of the body.

Accordingly, when people perfect themselves in character and spirit, the invisible love and deity of God appears through such a perfect person. This means that when perfect unity of mind (sungsang) and body (hyungsang) is achieved, then the perfect resemblance of God's sungsang and hyungsang is realized. The Bible calls the situation of the perfect man the sanctuary where God's love and deity make their appearance completely. In the Bible, the verse, "Surely you know that you are God's temple, and that God's spirit lives in you (Cor. 3: 16)!" speaks of this. All the religions and objects of religious salvation ultimately aim at unity of the mind and body so as to bring about a manifestation of the character of God's unity of sungsang and hyungsang. Hence, in the education of heart, we have to deal with educational methods to attain a man of character resembling God's unity of sungsang and hyungsang.

A) The Meaning of the Education of Heart

Education for bringing the individual to resemble God's perfection is education of heart. To resemble God's perfection is to resemble the unity of sungsang and hyungsang, which refers to the state in which the divine mind and physical mind engage in give-and-receive action centering on heart and become united in complete unity. Therefore, we need education to lead students to understand God's heart and experience it in their daily lives. This is the education of heart for the completion of individuals.

God's heart has been expressed in three forms through both the process of creation and the dispensation of restoration.[16] The first is the heart of hope. The heart of hope refers to God's joyful feelings, in anticipation of

---

[15] *Essentials of Unification Thought* (Japan; Unification Thought Institute, 1992), p.92

[16] Ibid., pp. 176-77

begetting Adam and Eve, His beloved children, to whom He could devote His ultimate love. God's expectation of joy was far greater than any kind of joy we may experience. In fact, the joy God felt when Adam and Eve were born was so great and deep that it cannot be compared to anything else.

The second is the heart of sorrow. The heart of sorrow is the Heart of God when Adam and Eve fell away from God into the realm of death under the rule of Satan. His grief at that moment was so great that there is no way to express it. When a child whom the parents love is dying, they feel desperate and grieve deeply. But there has been no one in history who has sorrowed more than God. This is one of the forms of God's Heart that Rev. Sun Myung Moon has taught us.

The third is the heart of suffering. God's Heart in the course of the dispensation of restoration, or, in the process of resurrecting fallen people, is the heart of pain or suffering. God's pain refers to the those feelings when He has seen his representatives, the saints and sages whom He sent, persecuted, imprisoned, and finally killed. Every time God saw the saints and sages suffering from persecution or imprisonment, He would feel as though a nail were being driven into His chest, or His side were being pierced by a spear. This is God's Heart of tears and regret.

Therefore, the education of heart is to lead the student to understand and feel this element of God's heart through the teacher. It is especially important to teach the sorrowful and painful heart of God which appeared in the process of restoration. In this case, the teacher must set an example through practice. Following the words and deeds of the teacher in daily life, students can experience the heart of God.

i) *Pestalozzi's Education of Motherhood and Heart*

In Values Education, for the individual to attain perfection of the first blessing, education of heart is required; for the family to attain perfection of the second blessing, education of norms is required; and for families, clans and nations to attain the perfection of dominion, the third blessing, education of dominion is required, including technical education, intellectual education, and physical education. No one of these three educational methods can be neglected. Until these three are carried out in balance, the ideal of education cannot be realized.

J.H. Pestalozzi (1741-1827) said that there are three fundamental forces forming human nature, namely, mental power, heart power, and technical power; these three, he reasoned, correspond to mind, heart, and hand. According to him, education of the mind is education of knowledge, education of the heart is moral and religious education, and education of the hand is the education of technique (including physical education). Among of these three type, Pestalozzi emphasized typically education of moral which is originated from heart, and defined that aim of education is to lead a man to the state of moral.[17] Further, he emphasized the importance of family relationships, which are the foundation of all human relationships and expand to society, nation, and all humankind.

The image of the ideal teacher advocated by Pestalozzi with a woman called "Gertrude," however, showed a tendency to attach more importance to moral education than to knowledge or technical education. Particularly, the three kinds of human nature he presented as the regulated principle of educational methods is poorly grounded to become a universal and permanent educational idea. We must know that the three forms of educational method in Unification Theory have a very strong logical foundation from the perspective of human nature or of the Bible. In short, education of heart is one of the educational methods for the perfection of the individual, through which sungsang and hyungsang can become united centering on Heart (Simjung) and come to resemble the perfection of God. Since the unity of sungsang and hyungsang is formed centering on simjung, it is the core of the education of heart for teachers (parents) to help students (children) understand and experience God's Heart.

Plato once called the mind which longs for Idea of the Good forgotten by people in the phenomenal world, the eros. Eros, in general, is human physical love. According to Plato, however, eros means the love seeking for and longing for the idea of beauty, wisdom and goodness forgotten by phenomenal man. In the Phaedrus, Eros is symbolized as wings to raise aloft that which is heavy to the divine region (beauty, wisdom, goodness) where the gods dwell.[18] Love, sensing a shortage of intellect from ignorance, is placed on a course seeking beauty within ugliness, and good surrounded by evil. Accordingly, the love advocated by Plato is not perfect, but a middle variety, the essence of which is to continue seeking for perfection. Eros ushers us on to the ultimate stage, that is, the sphere of the idea. Once it reaches that stage, however, eros finishes its role and must give way to pure intuition of the reason. This is because the sphere of ideas is the sphere of foreknowledge that can be understood by reason.

Therefore, the eros advocated by Plato is not essential, but only a middle and supportive function. According to Plato, the teacher should teach students with a mind of adoration and love for them and the matter of education is as important as giving birth to spiritual life. Plato, however, is biased toward rationalism and sees the object of education in rational ideas, outside the realm of erotic love. In unification Principle, "Heart is the emotional impulse to obtain joy through love," which is essential to God and humanity alike, and is the motive of creation. Hence, the education of heart in the Unification Principle is not supportive and dependent, but essential

---

[17] M.Liedke, J.H.Pestalozzi in *Selbstzeugnissen und Bild-dokumenten*,(Reinbek, 1968) pp.102-110

[18] F.M. Conford, *Plato's Cosmology*,(London Routledge & Kegan Paul,1971) p.354

and ultimate. In short, only through the education of heart, can the perfection of human character be fulfilled.

Heart is far deeper in meaning than agape which is the base of Pestalozzi's education and of Christian education. T. S. Nygren (1890-1970) makes a distinction between the nature of agape and eros, suggesting that eros belongs to Hellenism, while agape belongs to Hebraism. The love of agape, symbolized by the crucifixion of Jesus, means for people, through suffering death, to save others and realize their own true life. Agape is one's own sacrificial love. It was Pestalozzi who dealt with it as the core of educational thought. His educational thought is based on faith in God, which is, however, limited by the boundaries of human nature. For him, God doesn't work in man directly, but works for man through an agent called "Gertrude", an ideal mother with the love of agape. He, in the opening of Dust of a Hermit, proclaiming "Whether they live in a palace or under a thatched roof, they are all the same human beings," tried to set up an educational method out of pure love for humanity. Therefore, for him, the qualification of a teacher is sufficient with a mother's love based on religion, and the qualification of motherhood does not have to have any knowledge or refinement. As a result, mother's love as an educational method is forced to emphasize ideal moral education.

The method advocated by Pestalozzi to educate the love of humanity is very similar to the education of heart in Unification Thought. Pestalozzi's educational method is limited, however, to the sphere of the era, on the contrary, the education of heart in Unification Thought is based on God's Heart which is transcendental, and has manifested through the historical parents of humankind. The perfect embodiment of human character and humanity cannot be possible until we communicate with God's Heart which has manifested in the historical field as beloved parents. In Unification Theory of education, three forms of the historical Heart of God are claimed to be understood and experienced.

It was because of the heart of joy full of expectation and hope to meet His beloved children that God created the universe and all things over thousands of millions of years as the parents of humankind. Yet, the heart of such expectation and hope was, due to the fall of His children, transformed into the heart of limitless sorrow and wailing. This is the very God, the beloved parents of humankind with a heart of pain, who started the providence of salvation to save His Children who had fallen into the sphere of death, receiving all kinds of humiliation, persecution, and contempt throughout the process of history. Since these three kinds of heart with joy, sorrow, and pain have concretely appeared through the lives of the prophets and sages in the course of history in the actual world, how we should teach the contents to students is an assignment for the education of heart.

Contemporary education in our schools does not overcome the limits of society or the era. They are busy training students in the image of contemporary society. So the educational problems caused by these phenomena are so serious that we now see dangerous situations everywhere. This critical situation of education comes about because schools concentrate their efforts on simple knowledge and vocational education, training technicians merely, excluding the general principles of universal education, and neglecting morality and the holiness of humanity. In this critical situation, the Unification Theory of Education must persuasively present the permanent and universal theory of education and educational methods.

B) The Education of Norms

The education of norms refers to education to realize the perfection of the family. The education of norms is, therefore, education to resemble the harmony of God's masculinity and femininity, when man and woman become husband and wife, that is, education to obtain the qualification to become a spouse and form a family. The most important element in this is education on the sanctity and mystery of sex. Sex is something to be experienced only after marriage, and should never be experienced until that time nor can any deviation from chastity be permitted after marriage. Logically, therefore, the education of norms is education to promote a law-abiding spirit in students (children) who are to follow the way of Heaven and become principled beings. The education of norms should be practiced side by side with the education of heart. As norms regulate behavior, they tend to become formal or legalistic when they are not grounded in and infused with love.

i) Prerequisites for Education of Norms—The Resemblance between the Masculinity and Femininity of God

As mentioned above, education is the teacher's effort to bring up students to resemble God, and concretely, the process of leading them to resemble the divine image and character of God. Therefore, as explained, education nourishing the perfection of individuality through coming to resemble the divine character and image of God is the education of heart. Yet, it is necessary to explain more clearly the dual characteristics of masculinity and femininity, the attributes of the divine character and image of God in relation to the education of norms. Because coming to resemble God's masculinity and femininity means realizing the perfection of family, at the same time, the idea of the perfection of family is realized through the education of norms.

All things consist not only of the dual characteristics of sungsang and hyungsang but also of the dual characteristics of masculinity and femininity. Animals exist and multiply though male and female beings. Plants also exist and multiply through male and female organs. All creation including not only the biological world but also the physical world consists of positive characteristics and negative characteristics. Positive and negative ions, and protons and electrons in the atom are clear examples of this fact. In nature, when the sun sets, the moon rises, after hot weather comes to an end, cold weather sets in. According to the Bible, because it was not good for Adam to be by himself, God took a rib from Adam and made the woman Eve. Human beings are created in a pair system, so neither man nor woman can become a perfect human being without a spouse. This world is composed of masculinity and femininity because God, the ultimate causal being, exists as masculinity and femininity. Accordingly, God is the subject of both sungsang and hyungsang, being the subject of both masculinity and femininity.

In the book of Genesis, God created people in his own image: "So God created man in his own image, in the image of God created he him; male and female created he them (Gen.1:27)." "The image of God"—or "Imago Dei" in Latin—has been the subject of heated controversy among Christian theologians since the early church. Christian patristic theologians understood the image of God to be the rational essence, or spirit (nous) which only human beings have. This 'spirit' is understood as originally belonging to God and having been distributed to humans at their creation. This Christian dogma, understanding humanity in this way, however, is a typical example of the influence of Plato's philosophical perspective on humankind and has remained a controversial subject until today.

In Unification Principle, the "image of God" refers to man and woman together, as the Bible clearly says. The invisible God made His appearance through Adam and Eve, the first human beings created as God's children. The image of God was shaped into the image of neither animal nor plant, but humankind, and concretely embodied through the image of a man and woman. In this way, a man and woman who were created as substantial object appeared as the invisible image of God, and the focal point of the creation of the universe in which the invisible God was embodied and appeared. When the invisible God appears through the visible image of a man and woman, that is, God's substantial object, the idea of the universal creation can be realized. Accordingly, Unification Principle does not regard God as only the paternal god, God the Father, but presents a vision of God the Parents as the central figure of heaven and earth, because He is the God of the dual characteristics of masculinity and femininity. Since masculinity and femininity manifest differently in regards to the roles of subject and object, the masculine can represent both sexual natures. Therefore, in general, as the father represents the parents in the family, we can refer to God the Parents as God the Father or the Father God.

Feminist Theologians, who as part of the women's rights movement in the 1970s, argued that the concept of the male god—God the Father—taught by Christianity was a mirror of the unfair male-centered social structure. As we have seen the Bible represents the image of God as both man and woman. So we have the right to call God our mother. From the viewpoint of Unification Principle, the feminist opinion cannot be claimed to have no sound foundation, because God is the God of dual characteristics with the attributes of both masculinity and femininity. Nevertheless, if feminist theologians continue to advocate only the maternal god—God the Mother—then they come into collision with typical Christian theology that persists in arguing for an only masculine God—God the Father in the trinity. These two apparently conflicting opinions are actually complementary as seen in the Bible's theory of the Parental God.

The second of the Three Great Blessings—"multiply, and replenish the earth"—which God gave humankind is the word to establish family which resembles God's dual characteristics of masculinity and femininity. To establish family, norms and ethical orders are required in the relationships of family, that is, the relationship of parent and child centering on God, the relationship of husband and wife, the relationships of siblings, etc. Yet, the most fundamental of these are the ethics of the parent-child relationship and the husband-wife relationship. In a sense, education of heart can be said to be the education to experience the heart of the parental God - God the Parents - through the relationships of parent and child, while the education of norms is that to resemble God's masculinity and femininity through the relationship of husband and wife. Above all, judging from social problems such as the crisis of family identification, the merchandising of sex, the increase in the divorce rate, and juvenile delinquency, we are compelled to recognize that all these social problems are directly connected with the ethics and behavior of the family. Therefore, the education of norms is more important than anything else.

ii) The Meaning of the Education of Norms

The education of norms is education leading to the perfection of family where a man and woman become husband and wife and resemble the harmony of God's masculinity and femininity, that is, it is education to prepare students (children) to become qualified as original husband and wife. Since failing to keep the norm, God's Commandment, was the cause of the human fall, to learn to keep the Commandments of God is the education of norms.

Since humans were created with reason and law, that is, Logos, the education of norms means leading people to follow this reason and law, that is, the way of heaven. Therefore, the education of norms can be called the education of reason and law. The way of Heaven means the natural law and law of value working throughout the universe. Between these two laws, the law of value means norms. As there are vertical order and horizontal order in the universe, so there are in the family. Accordingly, in the family are the senses of both vertical value and horizontal value.[19]

As mentioned above, the education of norms should be practiced at the same time as education of heart, because love means to forgive and embrace all things, while on the contrary, norms themselves regulate strictly

---

[19] *Essentials of Unification Thought* (Japan, Unification Thought Institute, 1992), pp.154-55

and sometimes forcibly demand obligations only. Norms not based on heart easily become too formal, while love that is not constrained by norms too easily indulges blindly in pleasure. Hence, the education of norms should be practiced simultaneously together with education of heart in a homelike atmosphere.

C) Action Plan for the Education of Norms

One of the most important parts of the words spoken by Rev. Sun Myung Moon is contained in the teaching for the establishment of ethics for true family. In general, the family as a basic social unit is the place where a species is reproduced. In spite of the importance of family and the sex function within it, ethics in the western world places its focus on either individuals (liberalism) or nations (socialism). Christian Theology's approach to human salvation also rests on either individual salvation (typical theology) or social salvation (popular theology and liberation theology), ignoring the real importance of family. In sociology, family is treated merely as the sphere ruled by romance or irrational sex, and in Marxism, family is regarded only as a bourgeois organization. But in Unification Thought, family is the ultimate sphere for the perfection of humankind and the focus of God's providence of salvation. In brief, the ideal society and world is the expanding unit of the form of true family.

i) *The necessity of education for theism*

Unification thought is resolutely opposed to any theories or ideologies which treat humankind as purely physical beings or as an advanced species of animal evolved through "natural selection." Beginning with Aristotle (384-322 B.C.) who defined humans as social animals, such theories have manifested more recently in the Darwinian theory of evolution through survival of the fittest, in Freud's psychology, and in the materialism of Marxism-Leninism. These atheistic and material centered theories have deeply infiltrated into the field of public education under the guise of science education, and students are exposed to these thoughts in a defenseless situation. With their tremendous influence on current thinking, these theories have led to an increasing devaluation of human spiritual value and directly or indirectly to the modern crisis in ethics, morality, and the family. Because of such theories, people unconsciously assume that human beings are merely material forms evolved from animals, and thus true love has no soil in which to take root.

Charles Darwin (1809-1882) claimed that human beings had evolved by through natural selection - the "survival of the fittest" or the "law of the jungle" - and implicitly denied that humans were the creatures of love and norms created by God. Behind such theories lie the ethics of animal instinct and the law of the jungle as well as philosophies of hatred and struggle. The theory of the proletarian revolution of K. Marx(1818-1883), which could only be realized through a tremendous cost in blood, is nothing more than a philosophy of hatred and struggle for capturing wealth and labor from the moneyed classes. S. Freud (1856-1939) was deep influenced by Darwin, and taught that humans are animals controlled primarily by sexual impulses (libido). H. Marcuse (1898-1979), a left wing student of Freud's, insisted that eros should be freed from the limits of the moneyed classes' ideology so that erotic culture might be established. These theories have become the foundation to support and promote ideas of sexual freedom.

In the face of the erotic culture of sexual freedom advancing like storm waves, what shall we teach? The thought of Rev. Sun Myung Moon advocates Godism, showing strong disapproval of these atheistic forms of materialism. Godism puts the original cause of the holy spiritual value of humankind, morals and ethics, and the origin of the conscience in God, asserting emphatically that the spirit and body of human beings together with love and sex come from God and God alone.

ii) *Values Education prior to Sex Education*

As the mass media in general is full of contents depicting sex and violence, sex education itself has a tendency to become more sexual and stimulating. According to recent American research, those teenagers who received sex education had a higher rate of sex before marriage than those who did not receive sex education; 40% of 14-year-old juveniles, and 25% of 16-year-old juveniles showed a higher rate sexual activity than those who did not receive such "education." A family planning agency encourages using condoms and promotes sex education as a means of birth control. That some contraceptive device manufacturers and providers of surgical abortions enjoy a favorable business climate further illustrate this tendency.

Traditional sex education merely explained the differences between the physical bodies of men and women and went as far as is necessary. Human sexual appetite is not irresistible desire but an impulse to be controlled. This fact is not pointed out in contemporary sex education, and decadent ethics have spread due to such valueless education. Godism provides the educational methods to implant in youngsters true values concerning love and sex. From this viewpoint, the important element of the values education to be practiced in schools is follows:

First, the curriculum should introduce the necessity of a sense of values related to love and sex. According to Freud, love is "sexual love aimed at sexual intercourse." The loftiest human love is held to be locked up in sexual desire, because sexual energy (libido) is regarded as the motive power for all human behavior. But man consists of dual elements of spirit and body, that is, spirit body and physical body. Love belongs to the spirit body and sex belongs to the physical body.[20] As humans develop oneness between mind and body, human sexuality comes not to exist for sex itself, but

---

[20] Unification Thought Institute, *The Coming of the Age of Head-Wing Thought,* (Japan: Unification Thought Institute, 1997),pp.124-25

to realize love. Though animals retain sexual impulses coming from primitive instincts, human beings maintain sexual lives in the process of growing and maturing love. Sex is conjugal love expressed physically. Physical sexual love along with spiritual sexual love is absolutely necessary to realize conjugal love.

Second, the curriculum should introduce the necessity of a correct sense of values related to the dangers of free sex. Lasciviousness, in general, is said to be the love which takes place outside of conjugal relationships. Lascivious desire seeks to have sexual love with anyone, that is, it seeks free sex. According to Unification Principle, the root cause of lasciviousness is illicit love. Illicit love is the sexual impulse to use the object for the selfish pleasure of the senses. Sexual love from the libido in illicit love, that is, salaciousness, takes place with anyone other than one's spouse. According to St. Augustine (354-430), the original sin is the same as the passions, which means that aspect or tendency of the passions for physical pleasure to dominate reason and will. Hence, for Christians, marriage is needed as a means to calm down the passions to devote oneself to God. For Augustine, married life is a passive tactic to prevent the bigger sin, lasciviousness, through cooling off the passions.

In Unification Theology, sexual desire itself should not be exterminated as Augustine insisted; rather, it should be transferred from the fallen to the original libido, through which love in proper conjugal relationships can be restored. Sexual love is something absolute permitted only between husband and wife. Therefore, the cause of free sex is fallen sexual desire, which results in the breakdown of the true family. In human history, nothing has been more fearful than this sexual sin of adultery and free sex.

A man who has realized true love should love any woman besides his wife as mother, elder or younger sister, or daughter. In the same way, a woman who has realized true love should love any man besides her husband as father, elder or younger brother, or son. This is the essence of the education of norms according to Unification Thought. When this true love is developed and practiced, sexual desire for one's own physical pleasure cannot be triggered. This is the Movement of Absolute Values, and the Movement for True Family originated by the Unification Movement.

iii) The Movement for True Family and the Education of Norms

Since contemporary schools are slanted toward technical education, ignoring moral education, parents cannot but become teachers to take charge of this education. The education of norms and morals should not become compulsory or forced, saying "don't do this, don't do that." We should lead students to follow norms naturally while allowing them to do what they are willing to do. The family, therefore, must become a school of true love. In short, the education of norms should be practiced through true love. The root of morals and conscience is true love, and the root of true love is God. Hence, the foundation of ethics education should be set upon a theistic foundation, Godism.

Norms without true love are only forms, and love without norms and order is recklessness. Moral education should be sublimated not into an ethics of compulsion but one of pleasure. For growing children, compulsory ethics on the "Don't do that" principle do not have any persuasive power, because children are exposed to bigger, more stimulating temptations than the rules of parents or schools. What we need is a movement of true family. God's love is manifested in the prism of the family as children's love, siblings' love, conjugal love, and parents' love in separated characteristics."/> Through these separated loves the family is completed and God's love is manifested. The fundamental solution to juvenile delinquency, in conclusion, cannot be found without the ideals of true family.

In that sense, to educate according to the education of norms in Unification Thought can be regarded as the most positive way to solve the problems of young people. Today neither refined sex education in the advanced countries nor any traditional methods in Confucian countries have solved children's problems. It can be seen that no religion or philosophy can solve this worldwide problem of juvenile delinquency. To suggest an alternative idea, it is to rely on Unification Thought's approach to the education of norms as derived from Godism.

III. The Education of Dominion: intellectual, technical, and physical education

A) Education for the Perfection of Dominion

The education of dominion refers to education to perfect dominion. To perfect dominion, students must acquire requisite knowledge about the objects of dominion, both material and personal. This is education of knowledge, that is, intellectual training. The knowledge and study to exercise dominion varies according to the object of dominion; natural science for the knowledge of material objects, the fields of politics, economics, sociology, and the humanities for the knowledge of personal objects. In addition, the knowledge and technique for the development of creativity should be acquired. This is education of technique, that is, technical training. Education in the arts can be called a kind of education of technique. In addition to that, for education of Dominion, education for developing physical strength is needed, that is, physical training. Education of knowledge, edu-

---

[21] *Essentials of Unification Thought* (Japan; Unification Thought Institute, 1992), p.204

cation of technique, and physical training together we call education of dominion.[21]

B) Creativity and Education of Dominion

Education of dominion is to learn the means to exercise creativity. Creative potential is something with which everyone is naturally endowed, but education of dominion is needed for its practical exercise. Accordingly, as the activity of dominating objects through exercising creativity is "dominion," so creativity and dominion have an inseparable relationship. Next, let us look into the dual structure of creation,[22] along with the development of creativity, in relation to the education of dominion.

Creativity, in general, can be defined as "the ability to make new things." God's creativity, however, is understood as the nature or the ability of creation, which is not quite a proper understanding. God's creation was not an accidental occurrence, but came about through God's irresistible motivation and total objective intention. The creation is thus rooted in the motive of heart, that is, "heart motivation,"[23] where the internal and external four position foundations must be formed centering on the purpose of creation. Therefore, God's creativity can be defined concretely as "the ability to form the internal and external four position foundations centering on purpose." To explain this by comparison with the creative conduct of human beings, forming the internal four-position-base means planning (making blueprints), or developing new ideas, while forming the external four-position-base means that people make new productions in accordance with plans and blueprints using machines and materials.

Accordingly, to improve the ability to form external four-position-base, people must acquire a great deal of knowledge, and develop their ability to think and generate new ideas. From the viewpoint of the Theory of Unification Education, culturing the ability to form the internal four-position-base is knowledge education. In knowledge, there exist both knowledge to seek for the inner world through such disciplines as philosophy and ethics, but also knowledge of pure logic to search for scientific truth in the physical world. Yet all of this knowledge, in a broad sense, is related to the activity of dominion in developing creativity and dominating objects. Therefore, all internal and external knowledge becomes the object of knowledge education. On the other hand, to develop the ability to form the external four-position-base, we should acquire technique and increase the degree of skill. Education to meet this is technical education. As we easily come to know in the course of making a machine or building a house, technical education is education to create certain objects with knowledge gained

---

[21] Ibid., p.184

[22] Ibid., p.185

[23] Ibid., p.21

through the internal four-position foundation in order to realize a purpose, or, to cultivate dominion.

C) Education of Dominion based on Universal Education

Education of heart and the education of norms are called "universal education" because every person should learn them commonly. On the other hand, the education of dominion must be offered to people according to their abilities, interests, and desires. In this sense, it becomes "individual education." Education of dominion should be based on education of heart and the education of norms, and carried out along side each other, because knowledge education, technical education and physical education cannot be something sound or made fully manifest until they are based on heart and norms. Accordingly, on the primary base of universal education (education of heart and the education of norms), individual education in schools should be carried out simultaneously. It can be said that universal education and individual education are in the relationship of sungsang and hyungsang. When universal education, sungsang education, and individual education, hyungsang education, become harmonized, true "balanced education"[24] can be carried out.

One of the most serious problems in school education today is the problem of balanced education. As current society becomes more and more pluralistic, itemized, and expertised, the person this society needs is a technician trained and with special expertise in a special field. Therefore, as schools are forced to produce only vocational technicians to meet the needs of society, the proper academicism of college has disappeared already. In the field of education, the liberal arts have been weakened, and the study of universal values or ethics for humanity gives the impression of work which is hopelessly out of date. Education in schools is different from that in training centers or institutes for technicians because schools are the only place where education in humanism and in the liberal arts can be carried out. In the Unification Theory of Education, the sphere of individual education—education to acquire expert knowledge and vocational technique, which corresponds to education of dominion—is important. However, education of dominion, without the premise of universal education—education of heart and the education of norms—is certain to become blind education. This is why universal education and individual education should be carried out side by side; so that the field of education can become balanced and harmonious.

D) The Image of the Ideal Educated Person

Up to now there have been many kinds of education, each with its own image of the ideal person corresponding to its own idea of education. The

---

[24] Ibid., pp.186-87

Unification Theory of Education also has its image of the ideal person. The image of the ideal educated person in the Unification Theory of Education is as follows: first of all, a person of character; second, a good citizen; and third, a genius.[25] These are the images of ideal man and woman, corresponding, respectively, to the education of heart, education of norms, and education of dominion. Therefore, when education is seen in terms of the image of the ideal person, the education of heart may be called education to develop a person of character, the education of norms may be called education to develop good citizens, and the education of dominion may be called education to develop genius.

i) *The education of a Person of Character*

The person of character mentioned in Unification Thought is the image of the ideal person manifested through the education of heart. In general, a person with a certain degree of virtue, knowledge, and health is called a person of character, but in Unification Thought, a person of character is one who has internalized God's Heart and who can practice true love. The person of character is thus someone who comforts God's sorrow and suffering with his whole heart in a spirit of filial piety, and practices God's true love, forgiving even enemies, while He has righteous indignation toward them. In short, the person of character is one who practices God's true love for all people and all things. Therefore, the image of the ideal person of character is someone who has perfected the whole personality, having developed the faculties of intellect, emotion, and will in a balanced way, on the basis of heart.[26] Accordingly, education to become a person of character is to help a person internalize God's Heart and practice His true love in daily life and become a true person of character. Many religions describe the ideal nature of human beings as love, mercy, and goodwill, and carry out religious education related to those virtues. It can be a part of the education for the person of character to realize the image of the ideal person of Unification Thought.

ii) *The Education of a Good Citizen*

A good citizen, meaning a good member of a nation, is an image of the ideal person manifested through the education of norms. The education of norms may be given in schools, but the basis of it must be in the family. Since the family represents a miniature of the order of the universe, it can be said that society, nation, and world are expansions of the system of order in the family. In that respect, we can say that ethics has a deep relationship with order. In the family, ethics is a form of practicing love toward each of three objects from a certain position in the family-four-position-base. So, if ethics is to be brought into existence, that position should be decided in advance. Without order, ethics cannot be brought into existence.

---

[25] Ibid., p.187

[26] Ibid., p.188

One of the reasons for the breakdown of traditional values is that order in the current society is shaken or has collapsed. The most serious is that family order, the starting point of the order of society, has collapsed or is neglected.

As there is both vertical and horizontal order in the universe, so there is vertical and horizontal order in the family as well. The relationship connecting grandparents, parents, children, and grandchildren is the vertical order, while the conjugal relationship and sibling relationships are the horizontal order. Further, as the Logos maintains universal order, the Logos also maintains family order. The Logos operating in the universe is the way of Heaven, and the way of Heaven working in the family is family ethics. Social order, national order, and world order are resultant orders of the expansion and application of family order. As a result, a person who practices good education of norms in the family can maintain the normative life in society, nation, and the world as well, becoming a good member of society, a good citizen of the nation, and of the world, too. In other words, if a person becomes a good family member through the education of norms, he can behave properly, as occasion requires, for the norms of society, the nation, and the world wherever he may go. The education to foster humankind's good qualities and nurture students to manage normative lives for the good nation and world is the essence of the education of good citizens.

iii) The Education of Genius

Genius is the image of the ideal person to be formed through education of dominion. "Genius" in general refers to a person with special talent and ability, but the genius referred in Unification Thought differs from that in meaning. Originally everyone has the talent of a genius, since humans originally are beings of creativity, having been given God's creativity. Creativity is given to a person at birth as a potentiality. As the word "genius" itself means in Chinese characters "talent endowed by Heaven," so everyone is endowed with God's creativity from birth. Therefore, except for those who are mentally defective, all people can become geniuses as long as they manifest their creativity one hundred percent.

According to the Unification Theory of Education, a genius is a person who has manifested God's creativity through the education of dominion. Since human beings have been given individuality, if the creativity received is fully manifested according to their individualities, each of them may become a musical genius, a mathematical genius, a political genius or a business genius, respectively. But in fallen society, people cannot so fully display their creativity and most remain in mediocrity. In the face of the reality of the limits of education of dominion in current society, we feel keenly the necessity of the education to develop God-given creativity, on the base of Godism, so as to nurture geniuses, that is, creative beings.

REFERENCE

1. The Holy Spirit Association for the Unification of World Christianity, *The Divine Principle* (Seoul; Sung-Hwa Publishing Co. Ltd, 2004)
2. The Holy Spirit Association for the Unification of World Christianity, *Exposition of the Divine Principle,* (New York; HAS-UWC, 1996)
3. Unification Thought Institute, *Explaining Unification Thought* (Tokyo; Unification Thought Institute,1981)
--------*Essentials of Unification Thought; The Head-Wing Thought*, (Tokyo; Unification Thought Institute, 1992)
--------*The Coming of the Age of Head-Wing Thought*, (Japan: Unification Thought Institute, 1997)
4. J. J. Rousseau, tr. Barbara Foxley, *Emile* (London; L.M. Dent Sons Ltd., 1974)
5. Meg Parker, *Socrates and Athens*, (Macmillan Education Ltd. 1973)
6. Edit. Edith Hamilton Huntington Cairns, *Plato*, (Bollingen Series LXXI, Princeton. 1973)
7. Frederick Copleston, *History of Philosophy - Kant,* (The Newman Press, 1961)
8.*The Encyclopedia of Philosophy* vol.8, (Macmillan Publishing Co.1967)
6. F. Froebel, *The Education of Man,* (Clifton; Augustus M. Kelley, Publishers, 1974)
10. M. Liedke, J.H.Pestalozzi in *Selbstzeugnissen und Bild-dokumenten,* (Reinbek, 1968)
14. Charles Darwin, *On the Origin of Species,* (New York: Penguin Books, 1968)
15. Hebert Marcuse, *Eros and Civilization* (Boston: The Beacon Press, 1966)
16. International Religious Foundation, *World Scripture* (New York, N.Y.; Paragon House, 1991)

www.ingramcontent.com/pod-product-compliance
Lightning Source LLC
Chambersburg PA
CBHW032018170526
45157CB00002B/753